Integrated Water Management in Canada

This volume provides readers with an opportunity to learn from front-line water managers of watershed-based agencies across Canada about integrated water management (or integrated water resource management). In common with practice in much of the world, the responsibility for implementing integrated watershed management in Canada is fragmented. Each province and territory in Canada has developed unique approaches or governance models to guide decision-making in that regard. Thus, this edited volume enables readers from around the world to gain insight into the best practices in Canada for achieving success and addressing barriers to implement IWM.

Although there remains no consensus about how to "best" approach river basin management, some of the main observations include the following:

- There is a need to balance a focus on "the big picture", with scoping the scale and scope of planning activities in order that feasible and effective solutions can be implemented.
- Three types of integration are popular among the agencies included in the book: (i) among environment, economy and society; (ii) interactions between people and the environment; and (iii) integration (or coordination) of administrative activities.
- Much more attention is required to achieving effective engagement from Indigenous communities.

Most of the chapters were originally published in a special issue of the *International Journal of Water Resources Development*.

Dan Shrubsole is a Professor in the Department of Geography at the University of Western Ontario in London, Canada. His research focuses on water governance, with a particular interest on river basin planning and management, floodplain management, wetland management and diffuse-source pollution control. He has conducted research primarily in Canada and Australia. He is a past board member of the Canadian Water Resources Association and is currently President of the Canadian Association of Geographers.

Dan Walters is an Associate Professor in the Department of Geography at Nipissing University, North Bay, Canada and has been studying integrated watershed management in Ontario for 15 years. His research activities include assessing First Nations water risks, source water protection strategies, wetland conservation, nutrient management, cyanobacteria management, and the conservation authorities program.

Barbara Veale is Director of Planning and Watershed Management for the Halton Region Conservation Authority, a watershed management agency based in Burlington, Canada. Barb has extensive experience in implementing integrated watershed management, co-authored reports and journal articles on the subject, and provided advice to fledgling watershed management groups in Canada and elsewhere. Barb's doctoral research focused on watershed governance and explored the use of watershed report cards as decision tools for watershed management in Canada.

Bruce Mitchell, FRSC, is a Distinguished Professor Emeritus in the Department of Geography and Environmental Management at the University of Waterloo, Canada. His research focuses on water policy and governance, with particular attention to the implementation of integrated water management. He has conducted research in Canada, Australia, Britain, China, India, Indonesia, and Nigeria. He is a past President of the Canadian Water Resources Association, and an honorary Professor at five Chinese universities.

Routledge Special Issues on Water Policy and Governance
https://www.routledge.com/series/WATER

Edited by:

Cecilia Tortajada (IWPG) – Institute of Water Policy, Lee Kuan Yew School of Public Policy, NUS, Singapore

James Nickum (IWA) – International Water Resources Association, France

Most of the world's water problems, and their solutions, are directly related to policies and governance, both specific to water and in general. Two of the world's leading journals in this area, the International Journal of Water Resources Development and Water International (the official journal of the International Water Resources Association), contribute to this special issues series, aimed at disseminating new knowledge on the policy and governance of water resources to a very broad and diverse readership all over the world. The series should be of direct interest to all policy makers, professionals, and lay readers concerned with obtaining the latest perspectives on addressing the world's many water issues.

Routledge Special Issues on Water Policy and Governance

https://www.routledge.com/series/WATER
Edited by:
Cecilia Tortajada (IJWRD) – Institute of Water Policy, Lee Kuan Yew School of Public Policy, NUS, Singapore
James Nickum (WI) – International Water Resources Association, France

Most of the world's water problems, and their solutions, are directly related to policies and governance, both specific to water and in general. Two of the world's leading journals in this area, the *International Journal of Water Resources Development* and *Water International* (the official journal of the International Water Resources Association), contribute to this special issues series, aimed at disseminating new knowledge on the policy and governance of water resources to a very broad and diverse readership all over the world. The series should be of direct interest to all policy makers, professionals, and lay readers concerned with obtaining the latest perspectives on addressing the world's many water issues.

Transboundary Water Cooperation
Principles, Practice and Prospects for China and its Neighbours
Edited by Patricia Wouters, Huiping and James E. Nickum

Water Reuse Policies for Potable Use
Edited by Cecilia Tortajada and Choon Nam Ong

Hydrosocial Territories and Water Equity
Theory, Governance and Sites of Struggles
Edited by Rutgerd Boelens, Ben Crow, Jaime Hoogesteger, Flora E Lu, Erik Swynedouw and Jeroen Vos

Energy for Water
Regional Case Studies
Edited by Christopher Napoli

Legal Mechanisms for Water Resources in the Third Millennium
Select papers from the IWRA XIV and XV World Water Congresses
Edited by Marcella Nanni, Stefano Burchi, Ariella D'Andrea and Gabriel Eckstein

Integrated Water Management in Canada
The Experience of Watershed Agencies
Edited by Dan Shrubsole, Dan Walters, Barbara Veale and Bruce Mitchell

Integrated Water Management in Canada

The Experience of Watershed Agencies

Edited by

**Dan Shrubsole, Dan Walters,
Barbara Veale and Bruce Mitchell**

Taylor & Francis Group

LONDON AND NEW YORK

First published 2018
by Routledge

2 Park Square, Milton Park, Abingdon, Oxfordshire OX14 4RN
52 Vanderbilt Avenue, New York, NY 10017

Routledge is an imprint of the Taylor & Francis Group, an informa business

First issued in paperback 2020

British Library Cataloguing-in-Publication Data
A catalogue record for this book is available from the British Library

ISBN13: 978-1-138-58691-8 (hbk)
ISBN13: 978-0-367-58899-1 (pbk)

Typeset in Myriad Pro
by codeMantra

Publisher's Note
The publisher accepts responsibility for any inconsistencies that may have arisen
during the conversion of this book from journal articles to book chapters, namely
the possible inclusion of journal terminology.

Disclaimer
Every effort has been made to contact copyright holders for their permission to
reprint material in this book. The publishers would be grateful to hear from any
copyright holder who is not here acknowledged and will undertake to rectify any
errors or omissions in future editions of this book.

The editors of *Integrated Water Management in Canada: The Experience of Watershed Agencies* would like to dedicate this book in the memory of one of their respected contributors, Charley Worte, who sadly passed away on January 20, 2018.

Charley is remembered for his significant contribution to the conservation authority movement over many years, and to the evolution of watershed management in Ontario. When he was with Credit Valley Conservation, Charley led the development of the *Credit River Water Management Strategy* (1992), which was aimed at ensuring "abundant, safe and clean water, now and in the future for both the people and wildlife within the Credit River watershed" – the first of its kind in the province.

Charley also co-authored *Blueprint for Success: Restructuring Resources Management for Ontario* (1993), which promoted the importance of managing water on a watershed basis. In the early 2000s, Charley contributed to the conservation authorities' response to the Walkerton Inquiry, highlighting the importance of source water protection and the watershed approach.

While with Conservation Ontario, the provincial association which represents Ontario's conservation authorities, Charley led the conservation authorities through the source water protection program, from the post-Walkerton hearings to working groups to development of the legislation, and the creation of 19 source water protection plans. The Walkerton Inquiry (and related hearings) was established in June 2000 by the Government of Ontario in response to a tragic incident of bacterial contamination of the water supply in Walkerton, Ontario in May 2000. Just prior to his retirement, Charley also played a large role in developing a document entitled *Watershed Management Futures for Ontario* (2012), Conservation Ontario's whitepaper that helped to drive new conversations to modernize the provincial *Conservation Authorities Act*.

Charley has left a tremendous legacy. He is remembered by his colleagues as a friend and mentor; a source of honest and thoughtful advice; and a fearless, unflappable leader. Charley is survived by his wife Holly (Keizer) and two children, Heather and Ian.

Contents

CONTENTS

Citation Information

The chapters in this book were originally published in the *International Journal of Water Resources Development*, volume 33, issue 3 (May 2017). When citing this material, please use the original page numbering for each article, as follows:

Chapter 1
Integrated Water Resources Management in Canada: the experience of watershed agencies
Dan Shrubsole, Dan Walters, Barbara Veale and Bruce Mitchell
International Journal of Water Resources Development, volume 33, issue 3 (May 2017)
pp. 349–359

Chapter 2
Integrated watershed management and Ontario's conservation authorities
Charley Worte
International Journal of Water Resources Development, volume 33, issue 3 (May 2017)
pp. 360–374

Chapter 3
Implementing integrated water management: illustrations from the Grand River watershed
Barbara Veale and Sandra Cooke
International Journal of Water Resources Development, volume 33, issue 3 (May 2017)
pp. 375–392

Chapter 4
Lessons from implementing integrated water resource management: a case study of the North Bay-Mattawa Conservation Authority, Ontario
Paula Scott, Brian Tayler and Dan Walters
International Journal of Water Resources Development, volume 33, issue 3 (May 2017)
pp. 393–407

Chapter 5
Integrated water resource management and British Columbia's Okanagan Basin Water Board
Natalya Melnychuk, Nelson Jatel and Anna L. Warwick Sears
International Journal of Water Resources Development, volume 33, issue 3 (May 2017)
pp. 408–425

Chapter 6

The integrated watershed management planning experience in Manitoba: the local conservation district perspective
Colleen Cuvelier and Cliff Greenfield
International Journal of Water Resources Development, volume 33, issue 3 (May 2017)
pp. 426–440

Chapter 7

Applying integrated watershed management in Nova Scotia: a community-based perspective from the Clean Annapolis River Project
Levi Cliche and Lindsey Freeman
International Journal of Water Resources Development, volume 33, issue 3 (May 2017)
pp. 441–457

Chapter 8

Integrated watershed management in the Bow River basin, Alberta: experiences, challenges, and lessons learned
Judy Stewart and Mark Bennett
International Journal of Water Resources Development, volume 33, issue 3 (May 2017)
pp. 458–472

Chapter 9

The Northeast Avalon Atlantic Coastal Action Program: implementing integrated watershed management in Newfoundland and Labrador
Kailyn Burke
International Journal of Water Resources Development, volume 33, issue 3 (May 2017)
pp. 473–488

Chapter 10

Implementing integrated watershed management in Quebec: examples from the Saint John River Watershed Organization
Marie-Claude Leclerc and Michel Grégoire
International Journal of Water Resources Development, volume 33, issue 3 (May 2017)
pp. 489–506

Chapter 11

Setting the stage for IWRM: the case of the upper Kiskatinaw River, British Columbia
Reg C. Whiten
International Journal of Water Resources Development, pp. 1–15
https://doi.org/10.1080/07900627.2017.1419126

For any permission-related enquiries please visit:
http://www.tandfonline.com/page/help/permissions

Notes on Contributors

Mark Bennett is Executive Director of the Bow River Basin Council, Calgary Water Centre, Canada.

Kailyn Burke is part of the Northeast Avalon Atlantic Coastal Action Program, St. John's, Canada.

Levi Cliché is Executive Director of the Clean Annapolis River Project, Annapolis Royal, Canada.

Sandra Cooke is Senior Water Quality Supervisor at the Grand River Conservation Authority, Cambridge, Canada.

Colleen Cuvelier is District Manager at Little Saskatchewan River Conservation District, Oak River, Canada.

Lindsey Freeman is Program Manager at the Clean Annapolis River Project, Annapolis Royal, Canada.

Cliff Greenfield is Manager & Engineering Technologist at Pembina Valley Conservation District, Manitou, Canada.

Michel Grégoire is part of the Saint John River Watershed Organization, Canada.

Nelson Jatel is passionate about water. Living in Kelowna, Canada, he works at the Okanagan Basin Water Board, developing practical solutions that reflect the best available science, innovative policy and consensus approaches. Nelson has a background in freshwater science, a master's degree in Water Governance and teaches Applied Water Law at Okanagan College, Kelowna, Canada. Nelson practices Social Network Analysis with clients to help make 'invisible' networks visible.

Marie-Claude Leclerc is part of the Regroupement des organismes de bassin versants du Québec, Université Laval, Quebec City, Canada.

Natalya Melnychuk is a PhD student in the School of Environment, Resources, and Sustainability at the University of Waterloo, Canada.

Bruce Mitchell, FRSC, is a Distinguished Professor Emeritus in the Department of Geography and Environmental Management at the University of Waterloo, Canada. His research focuses on water policy and governance, with particular attention to the implementation of integrated water management. He has conducted research in Canada, Australia, Britain, China, India, Indonesia and Nigeria. He is a past President of the Canadian Water Resources Association and an Honorary Professor at five Chinese universities.

Paula Scott is Director of Planning & Development at North Bay-Mattawa Conservation Authority, Ontario, Canada.

Dan Shrubsole is a Professor in the Department of Geography at The University of Western Ontario, Canada. His research focuses on water governance, with a particular interest on river basin planning and management, floodplain management, wetland management and diffuse-source pollution control. He has conducted research primarily in Canada and Australia. He is a past Board Member of the Canadian Water Resources Association and is currently President of the Canadian Association of Geographers.

Judy Stewart is an Ambassador for the Cochrane Sustainability Plan, Canada.

Brian Tayler is Chief Administrative Officer, Secretary-Treasurer, at North Bay-Mattawa Conservation Authority, Ontario, Canada.

Barbara Veale is Director of Planning and Watershed Management for the Halton Region Conservation Authority, a watershed management agency based in Burlington, Canada. Barb has extensive experience in implementing integrated watershed management, co-authored reports and journal articles on the subject, and provided advice to fledgling watershed management groups in Canada and elsewhere. Barb's doctoral research focused on watershed governance and explored the use of watershed report cards as decision tools for watershed management in Canada.

Dan Walters is an Associate Professor in the Department of Geography at Nipissing University, North Bay, Canada and has been studying integrated watershed management in Ontario for 15 years. His research activities include assessing First Nations water risks, source water protection strategies, wetland conservation, nutrient management, cyanobacteria management and the conservation authorities program.

Anna L. Warwick Sears is responsible for the day-to-day operations of the Okanagan Basin Water Board including the collaborative water management initiative, water quality improvement programs and aquatic weed management. Anna has a background in population biology and watershed planning and was previously the Research Director for an environmental organization in Sonoma County, California, USA.

Reg C. Whiten is a professional agrologist and planning consultant with specialization in integrated watershed management and resource stewardship planning in northwest Canada. He has also served as Watershed Steward to the City of Dawson Creek (2010–2014) in British Columbia, and as Senior Planner to the Peel Watershed Planning Commission (2008–2009) in Yukon Territory. From his practice based at Moberly Lake, in north-east British Columbia, Reg's firm InterraPlan Inc. undertakes ongoing work with specialized focus on water source protection for communities, government and First Nations.

Charley Worte was part of Conservation Ontario, Newmarket, Canada.

Integrated Water Resources Management in Canada: the experience of watershed agencies

Dan Shrubsole, Dan Walters, Barbara Veale and Bruce Mitchell

ABSTRACT
Water agencies from 7 of the 10 Canadian provinces shared their experiences regarding history, successes, challenges and lessons learned with integrated watershed management. Based on these contributions, it is clear that an integrated approach does not mean 'all-encompassing'. Rather, it proposes desirable and feasible solutions through a systems approach based on sound technical information (e.g. biophysical and socio-economic), public engagement and monitoring. The roles of all participants must be clearly defined in order to promote success and facilitate implementation. Enduring and emerging challenges, such as adequate capacity and financing, engagement with Aboriginal communities and other stakeholders, and successful implementation, are identified.

Introduction

The idea for this special issue arose as a result of our 2014 article (co-authored with Charles Priddle) in the *International Journal of Water Resources Development*'s special issue on Revisiting Integrated Water Resources Management, which provided an overview of the 67-year history of Ontario's conservation authority programme (Mitchell, Priddle, Shrubsole, Veale, & Walters, 2014). As part of that research process, we invited water management practitioners from the conservation authority programme to present their experiences with river basin management to two special sessions at the June 2014 annual meeting of the Canadian Water Resources Association, in Hamilton, Ontario. Shortly after that meeting, we thought a more national perspective on the state of integrated water management (IWM) in Canada would be appropriate, and invited representatives of 13 water management agencies across Canada to reflect on and write about their experiences (i.e. history, structure, successes, challenges and lessons learned) with integrated water resource management (IWRM). While several contributions in this special issue are on the Ontario conservation authority programme, which reflects the focus of our 2014 article and the context outlined above, we are very pleased with the level of national coverage (7 of the 10 Canadian provinces) provided by this collection of manuscripts (Figure 1).

Figure 1. Location of Canadian water management case studies represented in this special issue.

In our 2014 article, we characterized IWRM as

> an ecosystem approach in which at least: (1) the catchment or river basin rather than an administrative or political unit is the management unit; (2) attention is directed to upstream–downstream, surface–groundwater and water quantity–quality interactions; (3) interconnections of water with other natural resources and the environment are considered; (4) environmental, economic and social aspects receive attention; and (5) stakeholders are actively engaged in planning, management and implementation to achieve an explicit vision, objectives and outcomes. (p. 460)

We also acknowledged that "moving from the ideals of IWRM to successful implementation can be challenging" (p. 460), a view shared by others (Beveridge & Monsees, 2012; Biswas, 2008; Blomquist & Schlager, 2005; Butterworth, Warner, Moriarity, Smits, & Batchelor, 2010; Molle, 2008). Addressing this gap prompted us to undertake this current special issue on the experience of watershed management agencies in Canada.

We believe that the Canadian experience can be instructive for many researchers and practitioners throughout the world. In common with practice in much of the world, the responsibility for implementing integrated watershed management in Canada is fragmented, and there is a need for water management agencies to foster partnerships, coordinate planning and management activities, engage stakeholders, secure funding, monitor and report on progress, and update and adapt plans when necessary. All the provinces and territories in Canada have developed unique approaches or governance models to guide decision making in that regard. Thus, this special issue will enable readers to gain insight on the best practices in Canada for achieving success or addressing barriers to implement IWM.

We recognize a variety of strategies for planning and managing water in Canada, and that much has been and can be learned from those on the 'front lines'. Although this approach of having front-line managers report has at least one potential limitation – it can be difficult to comment on one's own shortcomings and lessons learned – it has been used successfully in the past. Thirty-five years ago, various key front-line managers and practitioners had an opportunity to present their views in a symposium and subsequent publication. In 1981, a co-sponsored symposium on River Basin Management: Canadian Experiences led to a book with 27 chapters which reviewed regional, provincial and interprovincial approaches being used across the country (Mitchell & Gardner, 1983). At that time, five key challenges were identified:

- There did not appear to be any single correct or proper way to pursue river basin management.
- There was an urgent need to reduce the time to complete and implement plans.
- There was a need to broaden the focus from 'water' to include related land-based issues.
- While there was a recognition of the merits of public participation, the results had been disappointing.
- There was a need to improve communication between those involved in writing plans and those who must decide if, when and what specific recommended initiatives are to be implemented and funded. (Mitchell & Gardner, 1983, pp. 1–4).

Since that time, there have also been a book that provided an overview of federal and provincial/territorial initiatives to plan and implement 'sustainability' in water management, which had the essential characteristics of IWRM (Mitchell & Shrubsole, 1994), and two edited volumes that allowed practitioners and academics to provide insights on IWM in Canada, although neither provided complete national coverage (Shrubsole, 2004; Shrubsole & Mitchell, 1997). This current special issue provides an opportunity to identify progress related to the challenges identified at the 1981 symposium, as well as subsequent findings noted in the volumes above, and to identify emerging problems and solutions, as well as opportunities.

The following sections provide context for this volume of contributions by providing an overview of the concept of IWM, the context for water management in Canada, and general observations arising from the manuscripts.

The physical and human contexts that frame water management in Canada

At first glance, Canadians would appear to have few concerns over the management of water. The country's population of almost 36 million people is served by about 9% of the global runoff. Another water statistic is that about 20% of the world's total water supply is in Canada, while it has about 0.5% of the world's population (Environment Canada, 2012). Although these types of data suggest that the country should have an abundance of water, it has long been recognized that this is a myth (Foster & Sewell, 1981).

Canada is tied for third with Indonesia, the United States and China in receiving nearly 6.5% of the global renewable supply of freshwater annually (Sprague, 2007, p. 24). Over 60% of the renewable water supply in Canada drains into the Arctic and sub-Arctic regions, while 90% of the population lives within 300 km of its southern border with the United States. The annual precipitation is variable, ranging from over 2000 mm on the west coast to less than

500 mm in Saskatchewan. Parts of the nation have experienced water shortages and drought. McBean (2015) observed that there also has been an increase in the frequency of major flood events, particularly over the past 60 years, which reflects a complex interplay between global climate change, and increasing occupancy and flood damage potential on flood plains. Some of the major recent floods include Winnipeg, Manitoba (1997), Peterborough, Ontario (2004), and Calgary, Alberta (2014).

In general terms, the Conference Board of Canada (2014) ranked Canada as having the fourth-best water quality among the 17 OECD countries. Three major risks to Canada's water quality are: inadequate treatment of sewage waste; industrial effluent; and runoff of fertilizers from agricultural areas. Three relatively recent events in Canada have prompted governments to realize how quickly water problems can have tragic impacts on people. The first two were the contamination of water supply systems in Walkerton, Ontario (2000), and North Battleford, Saskatchewan (2001). Seven people died and over 2300 people became sick in Walkerton from the bacteria *Escherichia coli* O157:H7, and many thousands of residents in North Battleford became ill as a result of a parasite, *Cryptosporidium*. The general response from all governments has been to increase the scope (e.g. source water protection) and depth of water quality regulations. Cyanobacteria is a recent problem that is triggering the call for more integrated approaches to water management. A mix of drought, flooding and water quality concerns underlies all the watershed organizations. Some of the articles in this theme issue provide more details on the nature of the integrated water management responses.

The third event reflects, in large part, the nature and history of relations between Canada's Aboriginal peoples and Europeans and Canadian governments. In the context of water, the 'tip of the iceberg' occurred in 2005, when unacceptable levels of bacteria were found in the drinking water of the community of Kashechewan (on western James Bay in Ontario) and residents were not informed in a timely manner (there was a delay of over two days). Subsequent studies found that the quality of the drinking water in many Aboriginal communities was much more degraded than in other locations. For instance, a 2008 study by the Canadian Medical Association found that of over 1700 boil water advisories issued in Canada, the vast majority were in Aboriginal communities (Eggerton, 2008). This event further raised the public's awareness of the need to protect water supply. The federal government recently pledged to eliminate the need for boil water advisories in Aboriginal communities across Canada. This special issue reveals that there is further need to integrate Aboriginal peoples and their perspectives in watershed decisions, which may help the federal government achieve its goal regarding boil water advisories.

Aboriginal peoples, who comprise just over 4% of Canada's population, have lived in North America since at least 17,000 BP (and possibly as early as 50,000 BP) and often settled close to Canada's ocean and freshwater coasts, as well as its many rivers and streams (Mulrennan, 2015). The historical relationship between European settlers and Aboriginal people is complex and often associated with conflict, displacement, attempts at assimilation, and a preponderance of negative outcomes for Canada's first peoples. The tardiness in notifying the residents of Kashechewan of water contamination, while the unfortunate result of the failure to hook up a back-up chlorinator and the absence of an emergency paging system for local water operators (a standard amenity in most Canadian communities), illustrates in a modest manner some aspects of this relationship. This event, combined with more recent initiatives, such as formation of the Idle No More movement (http://www.idlenomore.ca/) and the findings and recommendations arising from the Truth and Reconciliation Commission (http://www.trc.ca/websites/trcinstitution/index.php?p=905), has made the public of Canada more aware of the need to improve relationships with Aboriginal peoples across a wide range of issues, including the management of water and related land resources.

The Canadian Constitution Act (1982) supports a federal approach to government and divides responsibilities for water and other resource management between federal and provincial governments. The federal government has responsibility for Aboriginal people, and, as noted above, it has been playing a more significant role, and there is a desire to significantly change and improve past arrangements. Provincial governments have substantial influence over water management because (1) they own most of the water resources in their provinces, (2) they have ownership of land, mineral and forest resources that impact water, (3) they have responsibility for civil and property rights, and (4) they control the formation and responsibilities of local governments, which recently have been playing a more significant role in water management (Cairns, 1987; Pearce, 1986). With these four legal realities, provinces have enacted legislation pertaining to matters of water supply and quality, irrigation, drainage, recreation and power. As will be seen in this special issue, it is the actions of the provinces (and often local government) that have legitimized and focused many of the activities of watershed management agencies.

While the provinces derive their major powers by exercising proprietary and legislative rights, federal water management responsibilities have exclusive legislative jurisdiction over navigation, inland and ocean fisheries, interprovincial works, trade and commerce, and international relations. The federal government also has sole jurisdiction over federal lands and water north of 60° latitude, until and unless agreements are negotiated with the territorial governments. The federal government has influenced the activities of provincial and local governments through its spending powers, and in this volume, this is seen to arise in the federal government's direct funding of watershed management agencies in Newfoundland and Labrador, and in Nova Scotia (Burke, 2016; Cliché & Freeman, 2016).

Commentary

In common with a conclusion arising from the 1981 River Basin Management: Canadian Experiences symposium, the contributions in this special issue indicate that "there remains no consensus as [to] the 'best' way to approach river basin management" (Mitchell & Gardner, 1983, p. 2). However, the principles of IWM appear to be a common element guiding watershed management in Canada. The special issue illustrates the diverse contexts, situations and experiences of implementing IWM across Canada, which we believe should make it of interest to a wide range of readers. In this section we use insights from these authors to provide some comments on the current state of IWM in Canada.

First, while there is a realization of the value in recognizing and understanding 'the big picture', there continues to be a need to focus on the most important water and related land resource problems confronting the residents of a watershed in order that adequate attention and resources (e.g. human, political, or financial) can be directed towards implementing custom-designed solutions. An integrated approach does not mean all-encompassing; several front-line workers suggest that prioritizing local issues helps reduce the time to complete plans and garner local support and involvement (e.g. Cliché & Freeman, 2016; Cuvelier & Greenfield, 2016). This is essential as citizens are increasingly being asked to play a role in the planning, implementation and monitoring stages (Burke, 2016; Veale & Cooke, 2016).

The data and information generated by professionals and citizens play a fundamental role in guiding the ranking of priorities and the identification and assessment of alternatives.

Second, monitoring the outcomes of implemented programmes and projects has become a common practice for Canadian watershed agencies, although this can be resource-intensive. Water monitoring programmes are resource-intensive activities often beyond the capacity of watershed authorities. Partnerships with other agencies and post-secondary institutions have, in some instances, provided funds and expertise to monitor conditions (Melnychuk, Jatel, & Warwick Sears, 2016). In other situations, standardized protocols allow citizen volunteers with sufficient training to fill this need (Cliché & Freeman, 2016; Veale & Cooke, 2016). The emphasis on monitoring and partnerships, particularly with the voluntary sector, is a new element to integrated water management since the 1981 conference. Although database management and GIS can support data collection and analysis, effectively and efficiently coordinating the monitoring activities of multiple sources can be a logistical challenge. It will also be interesting to observe how agencies continue to engage citizens in monitoring activities over the long term.

Third, the practice of integrated water management in Canada often involves developing a holistic perspective, and applying a systems approach that focuses attention on answering the central questions of what needs to be done, by whom, and with whom paying for planning and implementation. At least three levels or types of integration, particularly at the watershed scale, are considered:

- Integration of the linkages among environment, economy and society (e.g. sustainable development)
- Understanding resource interactions and how humans have affected or may in the future affect natural processes, often as they relate to one or more of the following: water quality and quantity; surface water and groundwater; water and related land resources; and how human activities have contributed or may contribute to degradation
- Coordinating the responses in the context of a programme and/or project(s) that involve decisions about the mix of means (e.g. information and education, technical assistance, financial incentives, regulations, taxation, property acquisition) to solve the problem(s), and division of costs and benefits.

A systematic planning process often guides this three-level integration process. The watershed organizations often play a coordinating and integrating role that is crucial for maintaining momentum to achieve established milestones or goals (Cuvelier & Greenfield, 2016; Veale & Cooke, 2016). The responsibility for maintaining this cyclical process is being devolved to local watershed authorities, under the direction of boards of directors, and reflects a significant change since the 1981 symposium.

Fourth, the watershed-based agencies are continuously striving to clearly define their role in (contributions to) solving water problems, and to maintain and ideally increase the level of confidence from the public as well as key decision makers (Worte, 2016; Leclerc & Grégoire, 2016). This represents a shift since the early 1980s. Communication between the planners and practitioners is better aligned through multi-stakeholder planning and implementation oversight (e.g. Cliché & Freeman, 2016; Stewart & Bennett, 2016). The watershed organizations often play a coordinating and integrating role. Developing effective partnerships with other relevant public agencies is now a common practice. This can include inviting representatives from these other agencies to participate in planning exercises. In this way, implementation can be fostered because there has been engagement from all participants about the nature of the problem(s), the need for action and who is best suited to implement solutions in a coordinated manner (Veale & Cooke, 2016). There is now more emphasis on sharing resources and responsibility for completing certain tasks in watershed plans. The emerging information can be used to form part of the budget process of public agencies.

Fifth, relative to the experience reported in 1981, the planning process is now relatively less complex and more appropriate in length relative to the nature of the problem to be solved. There is acute awareness of the need for planning to be able to transition quickly to implementation, and there is often a conscious effort to achieve short-term, visible gains that can be seen as a product of the process (Veale & Cooke, 2016). The previously mentioned monitoring programmes aid in communicating to the public and decision makers the outputs, outcomes and impacts of implementation. Watershed organizations' websites and government data portals make information more readily available to the public.

Sixth, public participation/engagement remains a crucial undertaking during planning and implementation with continuing devolution of responsibility to local government or groups to plan, implement, monitor and update watershed plans. In addition, since there is reasonable public support for the IWM activities described in this special issue, there appears to be more attention devoted to designing governance arrangements that are effective, efficient and fair. Public participation is now mandated in some jurisdictions, for example Ontario and Manitoba (Cuvelier & Greenfield, 2016; Worte, 2016). In other jurisdictions, such as Alberta and Quebec, the voluntary nature of the programme encourages public involvement and initiatives (Leclerc & Grégoire, 2016; Stewart & Bennett, 2016). Public involvement is also sought because strategies for resolving water problems involve soft solutions or behaviour changes. Education and outreach programmes are a key feature of many IWRM strategies (e.g. Scott, Tayler, & Walters, 2016; Veale & Cooke, 2016). While there has been progress since the 1981 symposium (better engagement and partnering with Aboriginal communities), it is one aspect of IWM that requires greater attention (Melnychuk et al., 2016; Scott et al., 2016).

Seventh, a variety of financial arrangements support all the activities of the watershed agencies. In Atlantic Canada, the federal government's Atlantic Coastal Action Program was and is fundamental to the activities occurring in the Avalon Peninsula and Annapolis Valley (Burke, 2016; Cliché & Freeman, 2016). Most other agencies report a mix of funding from provincial government agencies and self-generated revenue (e.g. Leclerc & Grégoire, 2016; Worte, 2016). None has the ability to tax individual property owners or levy income taxes. The level of funding varies considerably, reflecting the nature of responsibilities, and ability and willingness to pay. As evident in the special issue, all local watershed authorities must be prepared and able to adapt to the shifting priorities of senior levels of government. This is one of the continuing challenges.

Key current challenges

The Canadian experience in the past 35 years has addressed a number of the challenges identified at the 1981 symposium. The time to complete watershed management plans has improved, and there is more emphasis on implementation. There also is more emphasis on sharing resources and responsibility for certain tasks related to watershed plans (Stewart &

Bennett, 2016; Veale & Cooke, 2016). It is generally appreciated that an integrated approach is about building and maintaining relationships, and creating a sense of responsibility and accountability among partners. However, there also are enduring and emerging challenges.

First, droughts, floods and/or water quality concerns are what normally triggered the formation of each watershed organization featured in this special issue. Agricultural, urban and industrial intensification are making it difficult to manage these threats. A changing and uncertain future climate further complicates the task of watershed managers. Monitoring and reporting will be essential to track trends and identify emerging threats. The responsibility for monitoring is shared among different groups, such as government departments, citizen volunteers and post-secondary institutions. Ensuring quality control and quality assurance will require clear and explicit sampling protocols and training sessions (Burke, 2016). Solutions should be both science-based and socially accepted. However, integration of such data and information into decision making has often been slow or limited.

Second, capacity issues seem ubiquitous. While the full-time staff complement of the watershed organizations in this special issue ranges in size from two to over one hundred, they all cite financial, human, political and information challenges as limiting their capacity to tackle the complex socio-ecological issues. This concern is based on the desire and need to do more, and not an indication of limited success. Watershed organizations without stable core funding often seek project-based funding from a mix of sources to improve the health of the watershed. This need to be opportunistic and seek available funding from various sources often limits their ability to undertake long-term planning initiatives (Scott et al., 2016). Nevertheless, specific project-based initiatives contribute to the cumulative improvement of watershed conditions.

Third, there appears to be better recognition of the need to fund all aspects of the watershed planning process. However, implementation is challenging when the watershed organization has no legislative authority or legitimacy (Leclerc & Grégoire, 2016; Stewart & Bennett, 2016). In many instances, watershed authorities' programmes are subject to periodic review. For instance, a review of Ontario's Conservation Authority Act is currently underway to "improve the legislative, regulatory and policy framework that currently governs the operation and activities of the conservation authorities" (Ontario Environmental Registry, 2016). The provincial government sought broad public feedback on three central questions: (1) How well is the governance model working? (2) How are the programmes and services delivered by conservation authorities best financed? (3) What should be the role of conservation authorities in Ontario? There does not appear to be a direct effort to seek comments on how well programmes and services are being delivered. There were nearly 250 submissions from various interest groups, such as municipal, conservation authority, developer, environmental sector, public and Aboriginal representatives. These interest groups identified areas of improvement that were consistent with issues identified by the practitioners in this special issue, such as clarifying the mandate and regulatory authority of conservation authorities, encouraging sharing of information among partners, and updating and reviewing funding mechanisms. Based on an analysis of public responses, the provincial government identified five priorities for updating the Conservation Authority Act: "stronger oversight and accountability, clarity and consistency, updated funding mechanisms, collaboration and engagement, and future flexibility" (Ontario Environmental Registry, 2016).

How the provincial government chooses to address these five broadly defined priorities could present future opportunities for or challenges to implementing IWM.

Fourth, Aboriginal communities are often still not adequately consulted – for free and prior informed consent – across Canada. While there may be seats available at or invitations to meetings, Aboriginal perspectives are often not part of the planning, implementation, monitoring or adaptation processes. The problem is not one-way, however, because in some situations Aboriginal groups may decline to engage with a watershed authority, arguing that they should only interact directly with senior officials of the relevant provincial government. The jurisdictional issues are complex, and some continue to be resolved by court decisions. Aboriginal communities may not participate due to the uncertain legal implications of subsequent legislation or infringement of rights. However, there is much that can be done that does not require resolution of jurisdictional questions. The Truth and Reconciliation Commission calls upon all Canadian society to "renew or establish Treaty relationships based on principles of mutual recognition, mutual respect, and shared responsibility for maintaining those relationships into the future" (Truth & Reconciliation Canada, 2015, p. 326). IWM provides the opportunity for watershed authorities and Aboriginal communities to jointly develop new relations. Partnerships and collaborations thus underlie the ideal of integration.

There always will be scope and opportunity to improve capacity for integrated watershed management or IWRM, and it is unlikely that one standardized approach will ever be suitable in all situations. However, notwithstanding the reality of some significant challenges which need attention, experience with an integrated approach across Canada highlights that learning continues, and that improvements are steadily being made. Thus, perhaps the most basic lesson is that we need to maintain a willingness to monitor what we do, acknowledge when things do not work as anticipated or hoped for, continue to learn, and be willing to adapt and adjust from experience and new understanding. In that spirit, we hope that what has been learned from applying an integrated approach to water management in Canada over the last 35 years will be of interest and value to managers and researchers working in other countries across the world.

Acknowledgements

We wish to acknowledge and thank the many people who made this special issue possible. We are very grateful to all the contributors, particularly those employed in the various water management agencies. They have very responsible and busy job descriptions, and we thank them for making the time to write the initial draft and to follow up on required revisions and edits, and requests for additional information. This special issue would not have been possible without their cooperation. Karen Vankerkoerle, cartographer for the Department of Geography at the University of Western Ontario, worked very diligently in redrafting many of the figures and maps. We also thank Cecilia Tortajada, editor-in-chief of the *International Journal of Water Resources Development*, for her encouragement, constructive advice and support through the publishing process.

Disclosure statement

No potential conflict of interest was reported by the authors.

References

Beveridge, R., & Monsees, J. (2012). Bridging parallel discourses of integrated water resources management (IWRM): Institutional and political challenges in developed and developing countries. *Water International, 37*, 727–743.

Biswas, A. (2008). Integrated water resources management: Is it working? *International Journal of Water Resources Development, 24*, 5–22.

Blomquist, W., & Schlager, E. (2005). Political pitfalls of integrated watershed management. *Society and Natural Resources, 18*, 101–117.

Burke, K. (2016). Northeast Avalon Atlantic Coastal Action Program (NAACAP): Implementing integrated watershed management in Newfoundland and Labrador. *International Journal of Water Resources Development*. doi: 10.1080/07900627.2016.1238346

Butterworth, J., Warner, J., Moriarity, P., Smits, S., & Batchelor, C. (2010). Finding practical approaches to integrated water resources management. *Water Alternatives, 3*, 68–81.

Cairns, R. D. (1987). An economic assessment of the resource amendment. *Canadian Public Policy, XIII*, 504–514.

Canadian Constitution Act. (1982). Being Schedule B to the Canada Act 1982 (U.K.), 1982, c. 11.

Cliché, L., & Freeman, L. (2016). Applying integrated watershed management in Nova Scotia: a community-based perspective from the Clean Annapolis River Project. *International Journal of Water Resources Development*. doi: 10.1080/07900627.2016.1238344

Conference Board of Canada. (2014). *How Canada performs: Environment*. Retrieved April 15, 2015, from http://www.conferenceboard.ca/hcp/details/environment.aspx.

Cuvelier, C., & Greenfield, C. (2016). The integrated watershed management planning experience in Manitoba: The local conservation district perspective. *International Journal of Water Resources Development*. doi: 10.1080/07900627.2016.1217504

Eggerton, L. (2008). Despite federal promises, first nations' water problems persist. *Canadian Medical Association Journal, 178*(8), 985.

Environment Canada. (2012). *Water: Frequently asked questions*. Retrieved April 15, 2015, from https://www.ec.gc.ca/eau-water/default.asp?lang=En&n=1C100657-1#ws46B1DCCC

Foster, H. D., & Sewell, W. R. D. (1981). *Water: The emerging crisis in Canada*. Ottawa: Ontario: Minister of Supply and Services Canada.

Leclerc, M. C., & Grégoire, M. (2016). Implementing integrated watershed management in Quebec: Examples from the Saint John river watershed organization. *International Journal of Water Resources Development*. doi: 10.1080/07900627.2016.1251884

McBean, G. (2015). Climate change: Adapting to risks in a changing climate. In B. Mitchell (ed.), *Resource and environmental management in Canada (5th edition)* (pp. 195–220). Don Mills Ontario: Oxford University Press.

Melnychuk, N., Jatel, N., & Warwick Sears, A. L. (2016). Integrated water resource management and British Columbia's Okanagan Basin Water Board. *International Journal of Water Resources Development*. doi: 10.1080/07900627.2016.1214909

Mitchell, B., & Gardner, J. (1983). "Introduction". In B. Mitchell & J. Gardner (Eds.), *River basin management: Canadian experiences* (pp. 1–4). Waterloo, Ontario: Department of Geography Publication Series No. 20. University of Waterloo.

Mitchell, B., Priddle, C., Shrubsole, D., Veale, B., & Walters, D. (2014). Integrated water resource management: lessons from conservation authorities in Ontario, Canada. *International Journal of Water Resources Development., 30*, 460–474. doi: 10.1080/07900627.2013.876328

Mitchell, B., & Shrubsole, D. (1994). *Canadian water management: Visions for sustainability*. Cambridge, Ontario: Canadian Water Resources Association.

Molle, F. (2008). Nirvana concepts, narratives and policy models: Insights from the water sector. *Water Alternatives, 1*, 131–156.

Mulrennan, M. E. (2015). Aboriginal peoples in relation to resource and environmental management. In B. Mitchell (Ed.), *Resource and Environmental Management in Canada (5th edition)* (pp. 55–83). Don Mills Ontario: Oxford University Press.

Ontario Environmental Registry. (2016). *Conservation authorities act review discussion paper.* Retrieved May 20, 2016, from https://www.ebr.gov.on.ca/ERS-WEB-External/displaynoticecontent.do?noticeId=MTI1Mzgx&statusId=MTk0Mzk5

Pearce, P. H. (1986). Developments in Canada's water policy. In *The management of water resources: Proceedings from an International Seminar* (pp. 1–18). Toronto, Ontario: Institute for Research on Public Policy and Institute of Public Administration in Canada.

Scott, P., Tayler, B., & Walters, D. (2016). Lessons from implementing integrated water resource management: A case study of the North Bay-Mattawa Conservation Authority, Ontario. *International Journal of Water Resources Development.* doi: 10.1080/07900627.2016.1216830

Shrubsole, D. (Ed.). (2004). *Canadian perspectives on integrated water resources management.* Cambridge, Ontario: Canadian Water Resources Association.

Shrubsole, D., & Mitchell, B. (Eds.). (1997). *Practising sustainable water management: Canadian and international experiences.* Cambridge, Ontario: Canadian Water Resources Association.

Sprague, J. B. (2007). Great wet north? Canada's Myth of water abundance. In K. Bakker (Ed.), *Eau Canada: The future of Canada's water* (pp. 23–36). Vancouver, BC.: UBC Press.

Stewart, J., & Bennett, M. (2016). Integrated watershed management in the Bow River Basin, Alberta: Experiences, challenges, and lessons learned. *International Journal of Water Resources Development.* doi: 10.1080/07900627.2016.1238345

Truth and Reconciliation Canada. (2015). *Honouring the truth, reconciling for the future: Summary of the final report of the truth and reconciliation commission of Canada.* Winnipeg: Truth and Reconciliation Commission of Canada.

Veale, B., & Cooke, S. (2016). Implementing integrated water management: Illustrations from the Grand River watershed. *International Journal of Water Resources Development.* doi: 10.1080/07900627.2016.1217503

Worte, C. (2016). Integrated watershed management and Ontario's conservation authorities. *International Journal of Water Resources Development.* doi:10.1080/07900627.2016.1217403

Editors' Note

Dan Shrubsole, Dan Walters, Barbara Veale and Bruce Mitchell

In addition to the nine articles in the special theme issue of the *International Journal of Water Resources Development*, this book includes an additional chapter written by Reg Whiten, who has extensive first-hand experience in implementing IWM in the Upper Kiskatinaw River in Northeast British Columbia (Figure 1). First, the experiences in the Upper Kiskatinaw River provide further evidence of progress on the limitations outlined in the 1981 River Basin Management: Canadian Experiences symposium, such as improving water monitoring, prioritizing threats, and achieving more meaningful public engagement. Second, Whiten (2018) provides more context for experiences with implementing IWM in British Columbia, and among Aboriginal Peoples in Canada. Unlike the situation in the Okanagan Basin Water Board (Melnychuk et al., 2017), the Upper Kiskatinaw River does not have an official watershed designation under the provincial rules that established a higher level of oversight for development and habitat protection. However, both river basin management organizations lack the legal authority to enforce policy initiatives. Further, Whiten (2018) highlights the changing context for watershed management among Aboriginal Peoples in Canada. The source water protection plan in the headwaters (i.e. Bearhole Lake) of the Upper Kiskatinaw River requires annual reporting to the public and First Nations. Whiten (2018) also makes reference to the unanimous Supreme Court of Canada decision regarding Yukon Territorial Government's failure to uphold the Territories' Umbrella Final Agreement (1993) planning process. The Supreme Court ordered the entire planning process to start again. This decision demonstrates how the courts are slowly reconciling the rights of Aboriginal Peoples in Canada to fully participate in watershed management decisions.

References

Melnychuk, N., Jatel, N., & Warwick Sears, A. L. (2017). Integrated water resource management and British Columbia's Okanagan Basin Water Board. *International Journal of Water Resources Development, 33*(3), 408–425, doi:10.1080/07900627.2016.1214909.

Whitten, R. (2018). Setting the stage for IWRM: the case of the upper Kiskatinaw River, British Columbia. *International Journal of Water Resources Development*, pp. 1–15 https://doi.org/10.1080/07900627.2017.1419126

Integrated watershed management and Ontario's conservation authorities

Charley Worte

ABSTRACT

In Ontario, integrated watershed management has evolved into a fragmented, multi-agency environment that has made effective management difficult. In the 1990s, two approaches emerged – a local voluntary approach based on informal agency partnerships, and a regulatory approach established in provincial legislation. This paper describes the successes, challenges and lessons learned by drawing upon the experiences of Ontario's conservation authorities. Key lessons learned include the need for an interactive planning cycle and a multi-stakeholder decision-making process. While significant progress has been made in the practical application of integrated watershed management, significant challenges remain including the lack of a comprehensive policy and inadequate resources.

Introduction

Integrated water resources management (IWRM) is defined as "a process which promotes the coordinated development and management of water, land and related resources, in order to maximize the resultant economic and social welfare in an equitable manner without compromising the sustainability of vital ecosystems" (Global Water Partnership, 2000, p. 22). One of the ongoing challenges many jurisdictions face is how to translate the principles of IWRM into a practical approach to water management. Biswas (2008) noted that the concept of IWRM is so broad and all-encompassing that it is impossible to implement. The lack of operational guidance and objective evaluation criteria, and the complexity of the issues and interests, among other concerns, have led to a wide range of interpretations and little real progress in actual implementation.

To some degree, each jurisdiction develops its own unique approach to the management of water resources based on its geographic circumstances, its resource management challenges, and its legislative and institutional culture. It follows then that each jurisdiction will develop an interpretation of and and approach to implementing IWRM that are consistent with these circumstances.

At the time of its enactment in 1946, the Conservation Authorities Act represented a new approach to managing water in the province. Ontario's conservation authorities were founded on three explicit primary principles: watershed management unit; local initiative;

and partnership (Mitchell & Shrubsole, 1991). A number of circumstances and events in Ontario over the last 25 years have driven change in Ontario's watershed management approach. After decades of progress, rapid urbanization and to some extent agricultural intensification led to deteriorating watershed health; government funding constraints forced review of approaches and programmes; and the Walkerton water contamination tragedy forced a comprehensive review of drinking water safety programmes. Ontario experimented with two approaches to adapt the concept of IWRM, which in Ontario is referred to as inte-grated watershed management (IWM), into a workable approach to address Ontario's reality. The first was a bottom-up, voluntary approach exemplified by watershed strategies devel-oped by conservation authorities to address water issues, such as the Credit River Water Management Strategy (Credit Valley Conservation Authority, 1990). The second was a result of the drinking water contamination tragedy in Walkerton, Ontario, in 2000, which led to the development of a top-down, regulatory approach to deal exclusively with drinking water source protection (O'Connor, 2002a). Both of these approaches incorporate elements of IWM. This paper examines the evolution of these approaches, discusses their relative merits, and assesses the lessons learnt from Ontario's experience in implementing IWM.

Evolution of watershed management in Ontario

To provide the context for a comparison of the voluntary and regulatory IWM processes, it is helpful to understand the evolution of water resource management in Ontario by describ-ing the institutional arrangements and legislation.

While the term IWRM emerged around 1990, this was the rediscovery of a concept that is much older (Biswas, 2008). In Ontario the foundations of this approach go back more than 70 years. The historic settlement of Ontario led to the rapid conversion of a predominately forested landscape to agriculture, resulting in significant changes to the hydrologic regime. Subsequent urbanization exacerbated the impact, so that by the mid-twentieth century river systems suffered from severe flooding, soil erosion, drought and pollution. To address these problems, Ontario began experimenting with watershed-scale resource management planning in the Ganaraska River and Grand River watersheds. These experiments led to the establishment of watershed management agencies through the Conservation Authorities Act in 1946.

There are currently 36 conservation authorities, ranging from 500 km^2 to 7000 km^2 in watershed area (Figure 1). Most conservation authorities are in southern Ontario, where the Great Lakes created modestly sized watershed management units and where there was an adequate municipal tax base. Conservation authorities were established based on local initiative and are governed and largely funded by the municipalities in the watershed, the rationale being that local decision making would better reflect local circumstances and community needs (Richardson, 1974). Initially, funding for the conservation authorities' activities was based on a provincial–municipal partnership; however, the provincial com-mitment has diminished since 1990. While conservation authorities still receive project-specific funding from the province, the ongoing operations are now almost exclusively funded by the municipalities and self-generated revenue.

As conservation authorities came into existence from 1946 through the 1960s, it was the practice of the provincial government to undertake a broad watershed study for each newly established conservation authority, leading to a watershed plan that would be the basis of

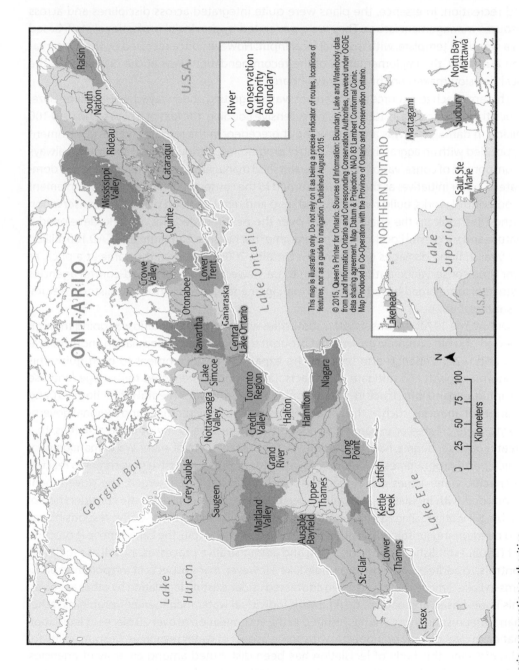

Figure 1. Ontario conservation authorities.

the authority's activities. For example, the *Credit Valley Conservation Report* of 1956 (Department of Planning and Development, 1956) was completed for the Credit Valley Conservation Authority, established in 1954. In addition to water management, these comprehensive plans provided recommendations on soil conservation, forestry, land use, wildlife and recreation. In essence, the plans were quite integrated across disciplines and across resource management issues. They were generated by provincial professional staff, based on a standard template, with minimal local input. However, once received by the conservation authority, the implementation of the recommendations was at the discretion of the local decision makers on the authority's board.

This approach to resource management was just getting underway when Hurricane Hazel struck the Toronto area in 1954, causing extensive flooding, the loss of 81 lives, and CAD 100 million in damages. It was an event without precedent in the region, and the government responded with an aggressive flood-control programme. Although flood control was always a component of conservation authority watershed management, it suddenly began to dominate all other initiatives as the province turned to the conservation authorities to implement flood-control works, build reservoirs, and map and regulate flood plains. The conservation authorities took up the challenge, and over the next few decades, Ontario developed a mature and successful flood-control programme that has saved an estimated CAD 100 million per year in flood damages (Conservation Ontario, 2009). However, this came at a price, as over time conservation authorities came to be viewed as primarily flood-control agencies instead of more broad-based watershed management agencies (Ontario Ministry of Natural Resources, 1987).

During the 1970s, as conservation authorities were maturing, Ontario was going through a period of rapid economic growth that brought the expansion of population, industrialization and urbanization. It also brought the expansion of government to both guide this economic growth and mitigate its impacts. Legislation was enacted to manage land use planning, expand infrastructure and mitigate pollution (Environmental Protection Act), and a variety of agencies, such as the Ontario Ministry of the Environment and regional governments, were created or expanded to administer these laws. Many of these activities, to a greater or lesser degree, had implications for the management of water. Water management policies and programmes in Ontario evolved over time as successive governments reacted to specific water issues and impacts. As each new problem emerged, the government responded with legislation and policy specific to that problem, with little consideration of the bigger water management picture. This led to fragmented administrative responsibilities and overlapping jurisdictions within the province. For example, the Lake Simcoe Protection Act (2008) established specific planning and administrative processes, as well as regulatory controls, to address pollution on a single lake. However, many aspects of the pollution problems in Lake Simcoe could have been addressed under existing legislation by other agencies. This issue-based approach of dealing with individual water management problems rather than water resource systems has resulted in the accumulation of over 20 pieces of legislation that each deal directly or indirectly with some aspect of water resources. Responsibility for administering this body of legislation has been distributed among an array of agencies, including local and regional municipalities, several provincial agencies and the conservation authority, with no formal process for coordination.

The result was a somewhat dysfunctional set of water management initiatives that often operated independently of each other, with no overriding strategic policy for the management of water resources as a whole. For example, while conservation authorities developed watershed plans for actions within their mandate, the Ministry of Natural Resources independently undertook fisheries management plans at a different geographic scale. Meanwhile, the Ministry of Environment regulated water withdrawals and contaminant discharges, without reference to either planning initiative. This fragmented approach to water management also created an institutional legacy of many agencies, ranging from local municipal governments to provincial government departments, each with responsibilities to fulfil and, to some extent, interests to protect.

As the population grew and land development and resource use intensified, the inconsistencies and conflicts due to the lack of coordination among agencies became problematic. For example, when dealing with urban development, the conservation authorities dealt with issues relating to flooding and erosion; the Ministry of Natural Resources dealt with issues relating to fisheries and natural habitats, such as wetlands; and the Ministry of Environment dealt with water quality issues. All of these interests led to requirements related to water that were not coordinated, leaving development proponents struggling to comply. By the 1980s, the limitations of this fragmented approach were becoming increasingly evident. Growing population and land use intensification were placing new demands on water resources, and the existing management structures were proving inadequate in mitigating the impacts. The fragmentation of responsibilities was creating a number of problems, including duplication of effort, conflicting requirements, and project delays (Ontario Ministry of Environment and Energy and Ontario Ministry of Natural Resources, 1994). The lack of a common vision or coordination of actions often created friction between agencies, and delays and confusion on the part of project proponents, stakeholders and the public. More important, however, was that in spite of these efforts this approach was proving ineffective, as water resources in urbanizing watersheds continued to deteriorate (Ontario Ministry of Environment and Energy and Ontario Ministry of Natural Resources, 1994). And while the issues impacting watersheds in more urbanized areas were complex, the lack of resources in rural watersheds created its own set of problems and inconsistencies across the conservation authority programme (Conservation Ontario, 2012).

Integrated watershed management

Eventually these problems reached the point where a few agencies began to search for a new approach. From 1990 onwards, several conservation authorities in the rapidly urbanizing areas of the province (e.g. the Greater Toronto Area) experimented with new approaches to water management in cooperation with their watershed municipalities and a few of the provincial agencies. This new generation of watershed plans focused primarily on anticipating and mitigating the hydrologic impacts of urban growth. While these watershed plans were a significant step forward, they were missing many elements of integrated water management, including key features such as the consideration of groundwater and aquatic habitat (Conservation Ontario, 2003).

The watershed management process continued to develop throughout the 1990s as conservation authorities experimented with expanding the scope and evolving the process of watershed planning, guided in part by the original watershed conservation plans and also influenced by the concept of IWRM, which was emerging internationally. This eventually led to guidelines for the development of watershed plans (Conservation Ontario, 2003) to promote better planning and to develop a more standardized approach. Three conservation authorities on the leading edge of this initiative, with cooperation from the Ontario Ministry of Natural Resources and Ministry of Environment, collaborated to document lessons learnt from their collective experiences and develop a practical approach for Ontario. The Toronto Region Conservation Authority and the Credit Valley Conservation Authority were both dealing with the impacts of exploding urban development in the Greater Toronto Area. The Grand River Conservation Authority was dealing with the challenge of urban expansion in a watershed constrained by water supply and assimilative capacity. Together, these three conservation authorities, which have a larger financial base than most, had extensive experience in watershed management, covering a diverse range of issues. This did not represent the experiences of the conservation authorities in general, where watershed issues varied and resources were often limited. What emerged from this collaboration was an integrated watershed-ecosystem approach that was adapted to Ontario's institutional structures and geographical circumstances. Its purpose was to develop a process of managing human activity and natural resources in a watershed context, reflecting environmental and social as well as economic benefits. It was based on planning, interagency partnership, public participation and adaptive management through ongoing monitoring and review. The watershed management framework outlined in the Conservation Ontario (2003) report became the basis of an IWM approach in the province. The Ontario IWM approach emphasized that it went beyond the management of water to include the relationship between water and land use. Conservation Ontario (2010, p. 11), the association of Ontario conservation authorities, defined IWM as

> the process of managing human activities and natural resources in an area defined by watershed boundaries. It is an evolving and continuous process through which decisions are made for the sustainable use, development, restoration and protection of ecosystem features, functions and linkages. IWM allows us to address multiple issues and objectives and enables us to plan within a very complex and uncertain environment.

Some of the key features of this framework include the interpretation of the concept of integration, the governance model used, and its adaptive structure.

One of the long-standing criticisms of IWM is that integration of all resources, all issues and all interests into one comprehensive initiative is simply impractical, due to the complexity and scope of such an undertaking (Biswas, 2008). This is the case in Ontario, due to the fragmented approach, the complexity of issues and limited resources. While a comprehensive approach was the initial goal of the conservation authority planning initiative, what eventually emerged was a multi-phase process with a limited scope focusing on specific issues or triggers. The concept of integration evolved to mean the collaboration of stakeholders and the coordination of actions, rather than a comprehensive consolidation. The pragmatic reality is that watershed planning initiatives are driven by a specific issue, such as a significant environmental problem, or a large-scale development proposal. The agency or agencies that are most affected by this issue are usually the ones that promote the need for the planning initiative and often provide the bulk of the resources. For example, a large-scale municipal land use planning initiative would trigger a study focused on implications of land development, funded primarily by the municipality. These issues or triggers then provide the focal point for the planning process and guide the scoping of the subsequent plan. However, instead of reverting to a single-issue water management study, the framework encourages collaboration and coordination through its governance structure and the adoption of a cyclical adaptive management approach. This interagency coordination and broader consultation facilitates identification of opportunities for synergies or potential conflicts with other water management issues and programmes. For example, the Rouge River Watershed Plan of 2007, while coordinated by the Toronto Region Conservation Authority, was completed by a task force comprising a wide range of public agencies and NGOs. It was triggered by the need to protect the ecological integrity of the park from encroaching urbanization and other associated impacts (Toronto Region Conservation Authority, 2007).

Perhaps the critical aspect of the Ontario approach is the governance model. In Ontario, most watershed plans have no legal status; they are undertaken voluntarily, and there is no obligation for anyone to comply with their recommendations. This has led to one of the principal philosophies of Ontario's IWM approach, which is that those who need to be involved in implementing the watershed plan also need to participate in the development of the plan, the rationale being that participation in the planning process is more likely to lead to support for its recommendations and ownership of their implementation responsibilities. Given the fragmented nature of the legislative basis of water management, this usually means that a number of Ontario's public agencies, as well as other stakeholder groups and the public, need to be involved in the planning process. Opening up the decision-making process to create a distributed governance model also addresses another criticism of IWRM (Biswas, 2008). Instead of water resource professionals being solely responsible for making water management decisions, other agencies and stakeholders bring the perspectives of diverse scientific disciplines, as well as non-science professionals. Scientific understanding is still critical to the process, but instead of scientists making the decisions, science informs the decisions made by the stakeholders.

The final feature is the adaptive management approach. As shown in Figure 2, the most significant feature of the framework is that it is a cyclical process. It recognizes that there are limitations to our understanding of natural and human systems, and to our ability to predict future changes, and uncertainty about the effectiveness of management actions. An adaptive management process allows for ongoing refinement of the plan over time. The watershed planning framework is based on a four-step sequence of planning, implementation, monitoring, evaluation (Conservation Ontario, 2003). While conceptually these steps occur

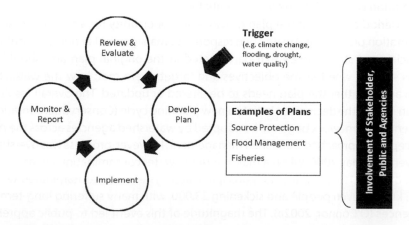

Figure 2. Integrated watershed management framework.

sequentially, in reality activity can occur at all steps concurrently, since, for example, management actions based on existing agency mandates do not shut down pending the development of a watershed plan. With these framework features as context, the following summarizes the IWM process.

The planning phase of the process is usually initiated by the conservation authority, but it is important to understand that the conservation authority is the coordinator and facilitator, not the sole decision maker. The coordinating role has proved successful due to its watershed jurisdiction and its governance structure. The conservation authority's watershed-based jurisdiction puts it in the position of having an overall understanding of the natural systems in the watershed and therefore being able to provide continuity among various watershed initiatives. Its structure of shared governance between watershed municipalities and partnership with the province puts the conservation authority in a fairly neutral position, balancing interests. Planning initiatives are usually directed by a steering committee comprising agencies with interests related to the issue at hand, and in some instances non-governmental agencies. Input from the broader stakeholder community and the public is sought at each step of the process through activities such as stakeholder committees and public meetings. The planning process itself follows a fairly conventional format, including characterization of the watershed system, goal setting, developing management alternatives, evaluating alternatives, selecting preferred alternatives, developing implementation plans, and monitoring and reporting (Conservation Ontario, 2003). Making the transition from IWM planning to implementation is a challenge in Ontario. There are a fair share of plans and strategies that have never made it off the bookshelf. Since there is no legislated requirement for IWM, implementation is not guaranteed but relies largely on the voluntary actions of the various agencies and stakeholders. This approach highlights the need for 'buy-in' from key partners and the public through the planning process and to develop a clear and detailed implementation plan that identifies specific agency responsibilities, required actions and expected timelines.

Ongoing monitoring is critical for a number of reasons. In the absence of a regulatory enforcement mechanism, monitoring and reporting progress on achieving objectives, actions and targets provides a means to track the implementation responsibilities of the various agencies. A comprehensive and integrated watershed monitoring programme also provides the data required to assess the effectiveness of management actions and expands the information base for future planning initiatives.

Finally, periodic review of the plan provides an opportunity for a formal assessment of implementation progress and watershed response to management actions. It should involve the agencies and stakeholders that participated in the original plan and should answer questions such as whether the objectives and targets were met; how the watershed has changed; and whether the plan needs to be altered or updated. The overall results of the evaluation inform the decision to initiate a new planning cycle (Conservation Ontario, 2003).

This generic framework began to be adopted by watershed agencies across the province as the accepted approach to watershed management. However, progress was overshadowed by the events of May 2000, when a major drinking water contamination event occurred in the town of Walkerton. The town's municipal water system was contaminated with *E. coli* O157:H7, killing seven people and sickening 23,000, with many suffering long-term health consequences (O'Connor, 2002a). The magnitude of this event led to public apprehension about the overall safety of water supplies and concerns about the capacity of government institutions to manage water. One of the consequences was that the existing efforts to develop and implement a broad-based IWM programme slowed as the provincial agencies focused attention on drinking water safety.

The government of Ontario established the Walkerton Commission of Inquiry, led by Justice O'Connor. The commission was charged with investigating the causes of the tragedy and making recommendations to ensure the future safety of drinking water across the province (O'Connor, 2002a). The process involved the broad participation of agencies, stakeholders and the public through formal submissions and consultations. Conservation authorities and other stakeholders saw this as an opportunity to establish a formal policy foundation for IWM; they advocated a broader, more integrated approach to the management of water, rather than a narrow focus on the protection of municipal drinking water sources.

Though there was resistance from some agencies and stakeholders to this broader, integrated response, this advocacy achieved some success. Justice O'Connor released two reports (O'Connor, 2002a, 2002b). The second made 117 recommendations, including endorsement of a comprehensive watershed management approach, although it did not make specific recommendations that would require it. It did, however, include a number of recommendations consistent with IWM. With reference to protecting drinking water sources, the recommendations included: watershed-based source protection plans be completed; planning be coordinated by conservation authorities as the existing watershed agencies; and affected stakeholders, agencies and the public participate in the development of source protection plans.

The provincial government embarked on an ambitious programme to implement the recommendations of this report. While there were a number of concurrent initiatives, it is the response to the recommendations for source water protection that most impacted the evolution of watershed management. The government established a Technical Experts Committee of agency and academic scientists, and an Implementation Committee of stakeholders and agency staff. These two committees were tasked with developing the policies and procedures required to implement a source water protection programme consistent with the inquiry's recommendations.

The committees continued to discuss the issue of developing a narrowly focused programme that dealt only with drinking water source protection versus a more integrated approach that embedded source protection within a more general IWM framework. While IWM offered the potential for more efficient and effective water management in the long term by dealing with a number of issues and interests concurrently, there were concerns about cost due to scope, time delays due to complexity, and perceived negative impacts on agency jurisdictions.

While IWM was not formally endorsed, the committee's recommendations did incorporate the concepts of watershed scale, agency cooperation and stakeholder participation. The work of these committees eventually led to the passage of the Clean Water Act (2006), which included a watershed-based approach coordinated by conservation authorities, with requirements for agency cooperation and formal public and stakeholder participation. For the first time in Ontario, watershed-based management was mandated by law rather than merely discretionary practice. However, the Clean Water Act focuses primarily on protecting drinking water sources.

Once the legislation was enacted, the government moved forward with an ambitious programme to complete source water protection plans in conservation authority jurisdictions. Under the direction of the provincial government, conservation authorities initiated the planning process by assembling source protection committees, which had the regulatory authority to produce source protection plans (Ontario Clean Water Act 2006, c. 22, s. 4 and 6). Source protection committees were stakeholder committees that included representation from the economic, environmental and social sectors. With conservation authority support, the committees completed the plans in two phases. First, watershed characterization and assessment reports assembled the existing watershed information and completed the analysis required to identify the various risks to drinking water sources. Second, risk management plans generated the policies necessary to eliminate or manage the identified threats and developed recommendations for implementing and monitoring plan policies. Each phase included a round of stakeholder and public consultation. The Clean Water Act included First Nations in the process, and by 2015 three First Nations communities had entered into agreements with the Ministry of Environment and Climate Change (Canadian Environmental Law Association, 2014). All 22 source protection plans were approved by the ministry before the end of 2015.

Comparison of voluntary and regulatory integrated watershed management

Ontario's experience with both voluntary and legislated approaches to IWM provides an opportunity to assess the relative merits of each. Implementation of source protection plans is still in its early stages, so a full evaluation is premature. However, based on the experience to date, a preliminary comparison is worthwhile. Many of the source water protection programme features are consistent with the voluntary IWM framework. Source water protection plans were watershed based, with conservation authorities acting as coordinators and facilitators. The governance structure was based on a steering committee comprising watershed municipalities and representatives from economic, environmental and social stakeholder groups. The source protection programme also followed the planning, implementation, monitoring, evaluation cycle. The Clean Water Act requires annual monitoring and reporting of the source protection plans (Sections 45, 46).

There are also important differences. Unlike the voluntary approach, the legislative backing of the Clean Water Act ensures participation, and because it was a government priority, adequate resources were available in all regions to complete the work. Provincial oversight through the review and approval of plans, in addition to the requirements for monitoring and reporting, strives to ensure a consistent approach and product.

Since it lacks a formal policy structure, voluntary IWM has the flexibility to adapt both the process and the content to incorporate the collective interests of the participating stakeholders, making it responsive to local concerns. This approach has the advantage of giving stakeholders and the local community the ability to influence the outcomes and have their priorities and values reflected in the proposed management actions. On the other hand, the mandatory approach comes with a rigid regulatory structure that imposes provincial values and priorities, which constrains both the process and product. For example, while the various stakeholder groups were well represented and public consultation was extensive, there was little flexibility in the process to address interests and concerns that extended beyond the government's priority to protect municipal drinking water sources. Many stakeholders brought concerns to the table, such as private drinking water sources, which could not be addressed.

The time required to complete the source water protection plans is of significant concern. It was 15 years from the Walkerton event, and 10 years from when source protection studies were initiated, to the final approval of the plans. Much of this time was taken up with developing specific legislation and regulations (seven years) and plan approval (three years). For comparison purposes, during this time the Grand River Conservation Authority undertook a water management plan with the voluntary approach, which covered a much wider range of issues, for a large and complicated watershed, that took only five years from plan initiation to final approval by all the partners. While many of the recommendations cannot be legally enforced, endorsement of the plan by the participating stakeholders signals their support for implementing the recommendations. It would appear, at least from this example, that an issue-specific legislative response takes much longer than a voluntary IWM plan. In addition, whereas the voluntary IWM approach gradually builds knowledge and institutional capacity to deal with a broad range of issues, specific legislative tools do little to enhance overall capability to deal with other problems (Worte, 2010). For example, the Clean Water Act can address only municipal drinking water sources. The Lake Simcoe Act targets only a specific geographic area: Lake Simcoe.

IWM in Ontario today

After nearly 70 years of experience, IWM is still very much a work in progress in Ontario. Like many other jurisdictions, Ontario has found it challenging to put the concept of integrated management into practice. Existing legislative and institutional arrangements, geographic circumstances and significant historical events have strongly influenced the evolution of today's IWM approach.

Since approximately 1990, Ontario has experimented with two main approaches to integrated management. The first to emerge was a bottom-up approach by the conservation authorities, which produced a generic IWM framework that provided a flexible adaptive approach and encouraged cooperation but was completely voluntary in terms of agency participation and implementation. The second was a top-down, legislative approach focused on drinking water source protection, which required participation and implementation but also imposed a rigid regulatory structure that did not always respond well to local circumstances or stakeholder interests.

The introductory article for this issue provides an overview of key principles of IWM, including: (1) the watershed as the management unit; (2) attention is directed to upstream–downstream, surface–groundwater and water quantity–quality interactions; (3) interconnections of water with other natural resources and the environment are considered; (4) environmental, economic and social aspects receive attention; and (5) stakeholders are actively engaged in planning, management and implementation to achieve an explicit vision, objectives and outcomes. Assessing the state of water management in Ontario, it is clear that, at least to some extent, all of these principles are being addressed by current practices.

First, a watershed management unit was established by the creation, in 1946, of conservation authorities as water management agencies with explicit watershed-scale jurisdiction. Operational experience since then, most recently through the source water protection programme, demonstrates that this management unit is respected during the planning and delivery of water management programmes.

Second, respecting the land-and-water connection in the watershed unit follows directly from fulfilling the first principle. Having an agency with a whole-watershed scale of responsibility leads to focusing attention on the interactions and linkages among the watershed components. This has been demonstrated in many watershed planning initiatives undertaken by conservation authorities over the last 25 years. Two examples include those completed by the Lake Simcoe Conservation Authority and the Grand River Conservation Authority, both of which have received the Thiess International Riverprize for best practices in river management.

Third, consideration of the relationship between water and other natural resources as well as the environment existed in the initial creation of the conservation authorities, and the early watershed conservation plans incorporated a range of natural resources as well as environmental issues. However, as mentioned earlier, this comprehensive approach drifted away as the conservation authorities' attentions became focused on flood control after the Hurricane Hazel flood event. As the IWM approach evolved from 1990 onwards, the relationships between water and other natural resources and environmental features became incorporated back into the voluntary IWM framework. However, in the source water protection programme, the regulatory structure kept the plans focused on drinking water and left little scope for broader environmental considerations.

Fourth, recognizing relationships among environment, economic and social aspects has been one of the most challenging, but considerable progress has been made in recent years. The early examples of water resources plans were initiated by water management agencies and staff with water management and natural resource expertise, and therefore focused primarily on the environmental aspects of water management. Economic and social aspects received little attention, since these interests were both beyond the mandates of the agencies participating and beyond the expertise of the individuals involved. Only in the last few decades have problems of increasing complexity forced the water management agencies to take a broader view (Ontario Ministry of Environment & Energy and Ontario Ministry of Natural Resources, 1994). Due to their limited mandate and expertise, conservation authorities started to involve agencies and stakeholders with economic and social interests in water management initiatives. This included watershed municipalities, which were interested in both economic development and the social interests of their citizens, but also other agencies and stakeholders. In essence, conservation authorities incorporated economic and social interests through the fifth principle, engaging stakeholders. The mandatory Clean Water Act approach was less effective at incorporating these relationships. Representation on the source protection committees by the economic, environmental and social sectors is built into legislation, but the constraints on the process resulting from detailed regulations and prescriptive directors rules limited actual consideration of these interests. For example, drinking water threats are defined in terms of human activities that can be a threat to drinking water sources (CWA, 2006, Reg. 287/07). Naturally occurring threats such as wildlife are beyond scope and cannot be dealt with.

The principle of engaging stakeholders is probably the most important aspect of IWM progress in Ontario. Fundamental to watershed management is the coordination of actions among those with responsibilities related to water. Given the historical fragmentation among a number of different agency mandates, it became clear that these agencies would have to be participants in any IWM initiative. Commitment to act by the various agencies needs to be preceded by participation in the planning process that determines the respective agency roles. In addition, land, at least in the southern part of the province, is primarily in private ownership. Given the linkages between land and water resources, water management initiatives may also affect land use, and therefore economic stakeholders and private landowners also need to be part of the planning process. In essence, the implementers of a watershed plan, be they public agencies or private landowners, need to be involved in the decision making that leads to the watershed plan. Both the voluntary and regulatory IWM approaches include stakeholders and the public in their respective processes.

Conclusion

Ontario has experimented extensively with implementing IWM, and both regulatory and voluntary approaches to IWM have yielded some successes. The voluntary conservation authority approach has been able to integrate issues at a watershed scale and has demonstrated an ability to engage stakeholders in the planning process and be responsive to local needs. The regulatory approach of the Clean Water Act has shown that some aspects of IWM can be implemented through legislation, can result in a more consistent process and can ensure stakeholder consultation in the process.

Some key lessons have emerged from the Ontario IWM experience. Due to the inherent complexity of natural systems, practical approaches to implementing IWM require a phased approach based on limited scope for specific initiatives. In the case of the regulatory approach, the scope is dictated by the legislation itself. In the voluntary approach, scope is driven by local priorities, which drive planning initiatives, and available resources. In both the voluntary and regulatory IWM approaches, the decision-making structure is key to successful implementation. Both approaches incorporate multiple agencies and stakeholders in the decision process, either regulated, such as the source committee structure, or informally, through the voluntary IWM approach. In either case, participation is expanded beyond traditional water management practitioners and agencies. In order to address the practical realities of limited resources, incomplete information and emerging issues, both approaches incorporate an integrative planning approach based on a planning, implementation, monitoring, evaluation cycle to allow for periodic refinement.

Despite considerable progress in developing and implementing IWM concepts, significant challenges remain. Ontario still has a fragmented legislative structure and lacks the comprehensive provincial water management strategy endorsed by Justice O'Connor (2002b) in the Walkerton report, Part 2. In the absence of the broad guidance of a provincial water strategy, integrated management still depends on the various agencies and stakeholders acting collectively on a voluntary basis, and water management agencies must continue to rely on informal local networks to coordinate activity. Progress in IWM in Ontario is also constrained by a lack of resources, particularly in more rural watersheds (Conservation Ontario, 2012). To date, water management tends to receive public attention and significant funding only during times of crisis, as demonstrated by the response to the Walkerton tragedy.

In an attempt to address this situation, conservation authorities, through their association, Conservation Ontario, have submitted a proposal to the government for a new approach to watershed management based on IWM (Water Management Futures for Ontario, 2012). This white paper proposes a discussion among the province, conservation authorities, municipalities and other stakeholders, which would achieve the following objectives: renewing and formalizing relationships between conservation authorities and various provincial ministries; refining the conservation authority model, including mandate, governance and accountability; and development of a sustainable funding formula (Conservation Ontario, 2012). In 2015, the Ontario government initiated a review of the Conservation Authorities Act. The purpose of the review is to "improve the legislative, regulatory and policy framework that currently governs the creation, operation and activities of conservation authorities that may be required in the face of a changing environment" (Ontario Ministry of Natural Resources and Forestry, 2015, p. 3). Discussions between conservation authorities and the provincial government are ongoing.

In the meantime, water management challenges are becoming ever more complex as the effects of more intense resource use and rapid urbanization are combined with the still-uncertain impacts of a changing climate. Degrading water quality in the Great Lakes, stormwater management issues from more intense flooding, drought, and urban water management problems are increasingly costly to address. Support for a shift to a more fully IWM approach would establish a process to address multiple issues concurrently to take advantage of potential synergies and avoid conflicts. This could enhance the efficiency and effectiveness of management interventions and build resilience on both a short-term and a long-term basis.

Although IWM has yet to be formally adopted in the province, it is firmly established in the initiatives of conservation authorities and within the limited scope of drinking water source protection planning. The provincial government's 2015 Provincial Plan Review of four key policy documents (Oak Ridges Moraine Conservation Plan, Niagara Escarpment Plan, Greenbelt Plan and Growth Plan for the Greater Golden Horseshoe) recommends that an integrated management approach be taken using the conservation authority watershed and sub-watershed plans (Crombie, 2015).

As new water management challenges emerge, water management agencies will have to find better ways to coordinate their actions to meet these challenges. Whether formally or informally, IWM is likely to continue to be part of that response.

References

Biswas, A. (2008). Integrated water resources management: Is it working? *International Journal of Water Resources Development, 24*, 5–22.

Canadian Environmental Law Association. (2014). *A first nations source protection tool kit*. Toronto: Author.

Conservation Ontario. (2003). *Watershed management in Ontario: Lessons learned and best practices*. Newmarket: Author.

Conservation Ontario. (2009). *Protecting people and property: A business case for investing in flood prevention and control*. Newmarket: Author.

Conservation Ontario. (2010). *Integrated watershed management: Navigating Ontario's future*. Newmarket: Author.

Conservation Ontario. (2012). *Watershed management futures for Ontario*. Newmarket: Author.

Credit Valley Conservation Authority. (1990). *Credit river water management strategy*. Mississauga: Author.

Crombie, D. (2015). *Planning for health, prosperity and growth in the greater golden Horseshoe: 2015-2041*. Toronto: Queen's Printer

Department of Planning and Development. (1956). *Credit valley conservation report*. Toronto: Queen's Printer.

Global Water Partnership. (2000). *Integrated water resources management: TAC background paper no. 4.* Stockholm: Author.

Mitchell, B., & Shrubsole, D. (1991). *Ontario conservation authorities: Myth and reality.* Waterloo: Department of Geography Publication Series No. 35, University of Waterloo.

O'Connor, D. R. (2002a). *Report on the walkerton inquiry: The events of May 2000 and related issues Part 1.* Toronto: Queen's Printer.

O'Connor, D. R. (2002b). *Report on the walkerton inquiry: A strategy for safe drinking water Part 2.* Toronto: Queen's Printer.

Ontario Ministry of Environment and Energy and Ontario Ministry of Natural Resources. (1994). *Water management on a watershed basis: Implementing an ecosystem approach.* Toronto: Queen's Printer.

Ontario Ministry of Natural Resources. (1987). *Review of the conservation authorities program.* Toronto: Queen's Printer.

Ontario Ministry of Natural Resources and Forestry. (2015). Discussion Paper: Conservation Authorities Act. Toronto: Queen's Printer.

Richardson, A. H. (1974). *Conservation by the people.* Toronto: University of Toronto Press.

Toronto Region Conservation Authority. (2007). *Rouge river watershed plan: Towards a healthy and sustainable future.* Toronto: Author.

Worte, C. (2010). Ten years after walkerton: Protecting municipal water sources. *Municipal World, 120,* 5–8.

Implementing integrated water management: illustrations from the Grand River watershed

Barbara Veale and Sandra Cooke

ABSTRACT

The Grand River watershed is the largest in southern Ontario. Poor water quality, floods and drought experienced in the 1930s prompted the formation of the Grand River Conservation Authority. While significant water improvements have been achieved, the Grand River faces chronic stress from the impacts of rapid population growth, land use intensification and changing climate. There is renewed commitment to address evolving water issues through integrated watershed management. This article summarizes the lessons learnt in the Grand River watershed and contends that integrated watershed management, although difficult to implement, provides a useful framework for practical application and positive results.

Introduction

The Global Water Partnership (2000, p. 22) defines integrated water resource management (IWRM) as 'a process which promotes the coordinated development and management of water, land and related resources in order to maximize the resultant economic and social welfare, paving the way towards sustainable development, in an equitable manner without compromising the sustainability of vital ecosystems'. The watershed as a management unit for water and related land resources underpins IWRM (Cortner & Moote, 2000; Goldstein & Huber-Lee, 2004; Heathcote, 2009; Saravanan, McDonald, & Mollinga, 2009; Mitchell, 2015). According to Shaver, Horner, Skupien, May, and Ridley (2007), addressing natural resource problems at a watershed scale rather than a single location or portion within it allows all relevant factors contributing to the problem to be included in the planning process, increasing the number of potential solutions to the problem or threat.

However, there are mixed views on the value of implementing IWRM. The application of IWRM for the day-to-day management of natural resources remains problematic, with significant gaps between theory and practice (Biswas, 2004; Blomquist & Schlager, 2005; Butterworth, Warner, Moriarity, Smits, & Batchelor, 2010; Mitchell, 2009; Molle, 2008). Some refer to IWRM as a 'nirvana concept' that promises more than it can deliver (Molle, 2008; see also Biswas, 2008). Others acknowledge that while IWRM has challenges, it is unreasonable to view it as a remedy for all environmental, social and economic woes. Rather, the principles inherent in the concept should be applied and the problems scoped based on context-specific factors. A watershed governance structure that allows inclusiveness, collaboration and accountability is one factor that will influence the success of IWRM (Plummer, Spiers, FitzGibbon, & Imhof, 2005). Another is the willingness of agencies to embrace an adaptive management approach, where lessons learnt are applied to future watershed planning initiatives (Veale, 2010).

The aim of this article is to describe the evolution and implementation of IWRM in the Grand River watershed in Ontario, Canada. A case study of these efforts will illustrate the successes, challenges and lessons learnt from experiences implementing IWRM. This remainder of the article is in four sections. First, the history and formation of the Grand River Conservation Authority (GRCA) are described. Second, the evolution of efforts to implement IWRM in the Grand River is illustrated by reviewing watershed planning initiatives. Third, the lessons learnt from implementing IWRM are considered: sustained collaboration; celebrating success; scoping the plan; and continuous improvement. These principles provide a useful framework to evaluate watershed and water management. The article concludes with a summary of the experiences implementing IWRM in the Grand River watershed.

Context

The Grand River watershed, the largest in southern Ontario, is located west of the Greater Toronto Area (Figure 1). The river emerges about 8 kilometres north-east of Dundalk and stretches almost 300 kilometres to the outlet at Lake Erie. Combined with its major tributaries, the Conestogo, Nith, Speed and Eramosa Rivers, the Grand River drains an area of 6800 km². The watershed is the largest of Ontario's drainage basins discharging into Lake Erie. Thirty-nine municipalities are wholly or partly within the Grand River watershed, as well as two First Nations reserves, formed under the federal Indian Act(1985).[1]

History of water management in the Grand River watershed

Water management has long been practised in the Grand River watershed. Key events over the course of the past century are summarized in Figure 2. In the 1800s, Europeans settled here, drawn by its abundant natural resources and fertile soils. The Grand River and its tributaries offered water supply, water power, transportation, and disposal for human and livestock waste. Over time, deforestation, agricultural drainage and urban settlement combined to intensify fluctuating river flows and contribute to poor water quality.

By the 1930s, floods, drought and pollution directly affected public health and economic development and prompted local business leaders to lobby the provincial government for action (Adams, 1937). The provincial government responded by authorizing a study, with clerical and financial support from municipalities (Mitchell & Shrubsole, 1992; Ontario Department of Lands and Forests, 1962). The study considered the problem of low river flow and its relationship to public health, water supply, sewage disposal, flood control and the provision of hydroelectric power. It recommended the construction of five dams and reservoirs, reforestation, wildlife management and the general improvement of the scenic features of the river valley (Ontario Department of Lands and Forests, 1932).

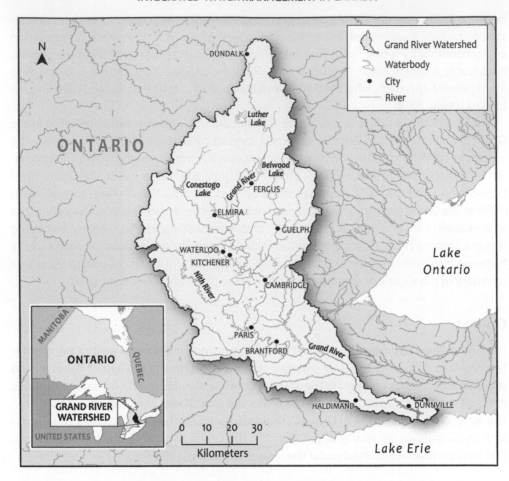

Figure 1. The Grand River Watershed, the largest watershed in southern Ontario, Canada.

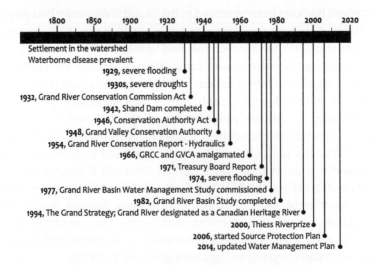

Figure 2. History of watershed management in the Grand River watershed.

The first watershed agency in Canada: the Grand River Conservation Commission

In 1932, the province established the Grand River Conservation Commission (GRCC) to undertake surveys; it was the first watershed agency in Canada. In 1938, after experiencing severe droughts, the province empowered the GRCC to investigate, construct and operate reservoirs. Members included the municipalities of Brantford, Elora, Fergus, Kitchener, Paris, Preston, Galt and Waterloo.[2]

The GRCC built three multi-purpose water control structures between 1942 and 1958, including the Shand, Conestogo and Luther Dams (Figure 1). The GRCC contributed 25% of the cost, and the remainder was shared equally between the federal and provincial governments. The GRCC also planted trees on lands adjacent to the newly created reservoirs.

In Ontario, natural resource concerns were not confined to the Grand River watershed. In 1946, the province passed the Conservation Authorities Act (1990), which enabled municipalities to collaboratively manage land and water resources on a watershed basis (Ontario Department of Lands and Forests, 1962). The Grand Valley Conservation Authority was formed in 1948. Activities were guided by a watershed-wide management plan. The plan focused on land acquisition, reforestation, local erosion and flood control, extension programmes for rural landowners, and recreational areas. It also encouraged all watershed municipalities to work together to address a broad range of resource management issues (Ontario Department of Lands and Forests, 1962).

Formation of the Grand River Conservation Authority

The practicality of two conservation organizations operating in the same watershed was assessed in the 1960s. As a result, the GRCC and the Grand Valley Conservation Authority amalgamated in 1966 to form the Grand River Conservation Authority (GRCA). The GRCA continued to carry out a wide range of conservation programmes and operate the multi-purpose dams and reservoirs.

In 1971, the efficacy of building more dams for water supply and effluent dilution to accommodate a growing population, versus the development of a pipeline to one of the Great Lakes, was debated and assessed (Ontario Treasury Board, 1971). As a result, the Grand River Implementation Committee was formed in 1972 to develop a comprehensive plan for the Grand River watershed.

In May 1974, a destructive flood caused an estimated CAD 7–10 million in damage (excluding business losses and clean-up costs) in the watershed (Grand River Disaster Relief Committee, 1975). The province established a Royal Commission to investigate the circumstances during and after the flood and offer advice to mitigate future damages. Improvements to the water control system, including a new dam and reservoir downstream from Fergus, were recommended (Leach, 1975). In 1979, a comprehensive environmental assessment of water control structures was completed by the GRCA. This study reinforced the call for an additional multi-purpose reservoir, together with river channel and waste disposal improvements (Grand River Conservation Authority, 1979).

Towards IWRM in the Grand River watershed

In 1977, the province initiated the Grand River Basin Water Management Study (Basin Study) to examine the interrelated issues of water quality, water supply and flooding. Directed by the Grand River Implementation Committee, this CAD 1.6 million study was led by five provincial ministries and the GRCA. The technical work was managed by a coordinator and undertaken by five multi-agency subcommittees. The study generated and examined 26 scenarios to deal with water issues over a 50-year horizon. Public input was sought throughout the process. Four public consultation groups, representing a wide range of interests and geographical areas of the watershed, helped form, screen and evaluate the scenarios.

The Basin Study was released in 1982. It offered 22 recommendations, calling for mix of structural and non-structural approaches to be undertaken by municipal and provincial governments and the GRCA. The anticipated cost of implementing the recommendations was more than CAD 180 million (Grand River Implementation Committee, 1982). Through the determined efforts of the partners, many of the recommendations were implemented, such as diking and the construction of sewage treatment plants. One key recommendation was to establish a coordinating committee to monitor implementation and consider periodic updates to the plan; however, this was never fulfilled (Project Team, Water Management Plan, 2014).

In 1987, another multi-agency initiative began which reinforced cooperative and coordinated management in the watershed. Several municipalities in the watershed were interested in pursuing Canadian Heritage River designation for the Grand River. A multi-agency steering committee, with representatives from Parks Canada, the Ontario Ministry of Natural Resources and the GRCA, was formed to explore whether the Grand River was a suitable candidate. After background studies (Nelson & O'Neill, 1989), the Canadian Heritage Rivers Board accepted the nomination of the Grand River in 1990, based on the watershed's rich diversity of cultural resources of national stature and the excellence of recreational opportunities provided by the river system.

Formal designation was contingent upon the tabling of a management plan with the board. A management plan called the Grand Strategy was produced after extensive public participation (Grand River Conservation Authority, 1994). A shared vision was debated, modified, and accepted early in the planning process. The principles of consensus, community involvement, cooperation and commitment underpinned the plan. Building on the philosophy that everyone sharing the resources of the watershed should be part of a cooperative effort to conserve, interpret and enhance river-related heritage resources, the motto 'Share the resources – share the responsibility' was adopted. The plan provided a framework for stakeholders to volunteer for specific actions (e.g. collecting and exchanging information, or assisting in implementing some projects) and was accepted by the Canadian Heritage Rivers Board in 1994, after which the Grand River and its major tributaries were formally designated.

In keeping with the vision and principles adopted for the Grand Strategy, several watershed-scale strategies and initiatives were promoted to address key aspects of watershed health and sustainability in the 1990s. An example is the Fisheries Management Plan.

Following the Canadian Heritage River designation in 1994, growing appreciation of the Grand River as a significant fishery resource and rising public interest in water quality and the environment led to the formation of a Grand River Fisheries Working Group. The purpose of the group was to investigate how angling interests and expectations could be met throughout the watershed. Representatives from angling groups, universities, provincial and federal agencies and the GRCA joined together to identify issues and potential solutions, advance stewardship and community sponsorship, interject fishery matters into land use and watershed planning decisions, and recommend management actions. Following extensive public input, the Grand River Fisheries Management Plan was published in September 1998 (Grand River Fisheries Management Plan Implementation Committee, 2005). Since that time, a multi-agency implementation committee has met regularly to carry out the recommendations of the report. In recognition of their continued work, the committee received the National Recreational Fishery Award from the federal Department of Fisheries and Oceans in May 2009 (Grand River Conservation Authority, 2009).

A significant event that influenced how water was managed in Ontario occurred in 1997. The province enacted the Water and Sewage Services Improvement Act (1997), which transferred ownership of wastewater treatment plants to municipalities (Ontario Sewer and Watermain Construction Association, 2001). This prompted some municipal water and wastewater managers to ask the GRCA to convene a working group to facilitate cross-municipal boundary discussions on water issues, including wastewater discharges to the river, non-point source pollution, water supply, and spill notifications. An early outcome of the discussions was the initiation of a Rural Water Quality Program, fashioned after a former provincial programme called Clean Up Rural Beaches (Project Team, Water Management Plan, 2014). Participating municipalities provided financial assistance to farmers to adopt best management practices, such as fencing near waterways and no-till farming, to protect and improve water quality. The Rural Water Quality Program continues today. The GRCA provides technical assistance and administers the programme. Over 4000 projects have been completed since the programme began (Project Team, Water Management Plan, 2014).

The long-term efforts of the GRCA and its partners have resulted in visible improvements to the river system. The multi-purpose reservoirs have successfully mitigated flooding and augmented low river flows, improving downstream water quality and ensuring more consistent flows for municipal water supplies. Upgrades to wastewater treatment and rural landowner stewardship projects have contributed to improvements in water quality and aquatic health, validated by the return of sport fish to several river reaches (Grand River Conservation Authority, 1997). The merits of this ongoing integrated and collaborative approach for managing water and related land resources were acknowledged in 2000, when the GRCA received the Thiess International Riverprize honouring excellence in river management (Krause, Smith, Veale, & Murray, 2001).

Updating the Basin Study: a renewed approach

Today, close to one million people reside in the watershed, most in urban areas. There are 45 municipal and one First Nation drinking water systems in the watershed.[3] Over 70% of the drinking water supplies come from groundwater sources; surface water sources account for the rest. There are four surface water intakes in the Grand River system. The region of Waterloo and the city of Guelph supplement groundwater drinking water supplies with surface water withdrawals, whereas the city of Brantford and the Six Nations of the Grand River depend solely on surface water withdrawals from the Grand River for their water supply. The Grand River receives treated effluent from 30 wastewater treatment plants. Also, non-point pollution sources, such as farms and urban development, contribute contaminants to the river (Project Team, Water Management Plan, 2014). The river also provides excellent opportunities for recreational fishing and paddling.

Despite continuing management efforts, the river system remains sensitive to stresses from rapid population growth, agricultural and urban intensification, and changing climate. The Grand River and most of its larger tributaries are highly eutrophic. Surface water quality is deteriorating in the middle and lower reaches of the river (Cooke, 2006; Loomer & Cooke, 2011). The capacity of the river system to receive additional wastewater and runoff from non-point sources is uncertain. The cumulative effects of nutrient and groundwater inputs are not well characterized or understood. There is increasing concern that the cumulative impact of progressive urbanization may negatively influence the water budget and disrupt fundamental hydrological processes, particularly in the central moraines area (Veale, Cooke, Zwiers, & Neumann, 2014). Without careful planning, the mounting demands on the river system could reduce the availability of water, increase water demand, and boost contaminant loads in surface and groundwater.

In 2008, the GRCA initiated dialogue with potential partners on the need to revisit the 1982 Basin Study to deal with existing and emerging water problems in the watershed. Since water management activities are undertaken by multiple agencies at three government levels, a business case was developed. There were two central messages. First, the resolution of water issues requires a collaborative approach that recognizes the complexity and inter-relatedness of hydrological, ecological and social systems. Second, solutions to address the impacts of multiple inputs and sources throughout the river system must be watershed based.

In 2009, the Grand River Water Management Plan (WMP) was launched to update the Basin Study. The approach was crafted based on previous experience with watershed-scale plans. Partners included municipalities,[4] provincial ministries, federal departments, First Nations and the GRCA (GRCA, 2014). The goals of the plan were to: ensure water supplies for communities, economies and ecosystems; improve water quality to improve river health and reduce the river's impact on Lake Erie; reduce flood damage potential; and increase resiliency to deal with climate change. These goals are broader than those of the Basin Study. While water quality, water supply and flooding were still considered top management priorities, a broader ecosystem-based philosophy was adopted, which acknowledged the impact of human activities on the health and functioning of natural systems. The influences of water quality and quantity on the ecological and hydrological health of the Grand River and Lake Erie, and the implications of climate change, were not considered in the Basin Study.

A Project Charter outlining the purpose, scope and partner roles was crafted and voluntarily signed by the partners, committing them to work together towards meeting the stated goals (Grand River Conservation Authority, 2010). The Project Charter stressed that the updated WMP would represent a 'joint call to action' by aligning the efforts of all partners and galvanizing them to achieve mutually supported targets for water management. The WMP was to be a collective and concerted effort to stretch limited dollars in support of 'best value' actions to maintain river health and resilience.[5] The plan was to build on past and current information, knowledge and modelling, planned and proposed projects and activities, and commitments to meet shared goals. It was not to be a prescriptive plan, offering recommendations for projects and programmes that ought to be undertaken. Rather, it was to represent the collective commitment of all partners to determined action, with each agreeing to implement specific actions.

The Project Charter outlined factors for success that helped guide the planning process. Some key factors included: (1) there is sustained collaboration among partners; (2) there are early wins to celebrate and share; (3) the plan is scoped to reflect available time, funding, resources, data and science; and (4) monitoring and performance measures are created to ensure adaptive management and continuity.

A multi-agency Steering Committee was formed, supported by a technical Project Team with representation from all partners. Several multi-agency working groups were also formed to answer technical questions, synthesize information, share best practices and exchange perspectives. The GRCA provided administrative support and ensured that information was shared freely among and discussed by participants. Funding for the plan was shared by the GRCA and the Ministry of the Environment and Climate Change, which provided CAD 903,000 through the Showcasing Water Innovation Program under the Water Opportunities Act (2010). Environment Canada also provided funding to help facilitate the planning process and ensure that links between the WMP and the Lake Erie Lakewide Action and Management Plan were considered.[6]

Early in the planning process, the partners adopted the vision created in the Heritage River designation process and the guiding principles derived from the Project Charter (Table 1). Twenty-three broad water objectives, supporting human uses, ecological needs and social and cultural values for water, were developed with broad stakeholder input and accepted by the partners. Indicators and targets describing conditions when the broad objectives are met were also identified.

The GRCA facilitated workshops, surveys and meetings with participants to gain knowledge of their experiences or insights on innovative approaches to achieving the water objectives. A series of best practices were developed to support improving the performance of wastewater treatment plants, improving stormwater management and applying innovative decision support systems. Further, demand management strategies were established for water withdrawals, such as municipal and agricultural users. An arms-length Science Advisory Committee, made up of highly respected academics with expertise in various aspects of water management, provided the Steering Committee, Project Team and GRCA staff with scientific, technical and management advice.

The WMP was completed and endorsed by 15 plan partners in September 2014. It is an integrated action plan, containing 163 actions. These actions include existing and planned activities, strategies, best practices and innovative approaches, which the partners consider practical, achievable and cost-effective. In 2015, the partners transitioned into the implementation phase. A renewed Water Managers Working Group of municipal, provincial and federal agency representatives, First Nations, and the GRCA meets quarterly to evaluate the progress each partner has made towards implementing the plan and to discuss ongoing or emerging water management issues. The first report, released in 2015, concluded that partners were advancing 120 of the 163 actions listed in the WMP. Initiatives such as a Drought Contingency Plan and the Grand River Source Protection Plan, the commissioning of a new water treatment plant at the Six Nations reserve, and the naturalization of 1.3 kilometres of Schneider Creek to improve urban stream water quality in Kitchener were highlighted, while other actions were noted as pending (Grand River Conservation Authority, 2015).

The WMP is one component of the integrated watershed management plan for the Grand River watershed, as illustrated in Figure 3. A planning process to tie all the individual watershed-based plans together to create an integrated watershed plan is slated to begin in 2016.

Table 1. Guiding principles for partners to the update of the Grand River Water Management Plan.

Healthy communities and a healthy ecosystem
- A healthy river system is crucial for sustaining prosperity, growth and well-being in the Grand River watershed
- The Water Management Plan is guided by an ecosystem approach. We will strive to maintain and restore critical natural system interactions, functions and resiliency
- Ecosystem services (those services provided by natural processes, e.g. waste assimilation, water retention) are acknowledged, maintained and enhanced

Managing water resources is a shared responsibility
- Managing water requires common goals, collaborative decision making and co-operation
- Implementation is shared by all levels of government, landowners, businesses and residents
- Implementers are committed to joint action and own their piece of the Plan
- Stakeholder participation is essential

Water is best managed on a watershed basis
- The watershed is the most appropriate unit for managing water and the linkages between water and other natural resources
- The Water Management Plan is a critical component of a broader watershed management plan

Decision making must be transparent and responsive
- Water management decisions are integrated and transparent, taking into consideration the broad range of uses, needs and values for water and the needs of a healthy ecosystem
- Water management strategies are designed to be responsive to changing conditions, priorities, vulnerabilities and pressures; adaptation is supported by monitoring and progress reporting

Management of water resources must be effective and efficient
- The concepts of sustainability, adaptive management and continuous improvement guide decision making and implementation
- Best value solutions are sought
- Best available science, expert advice and local knowledge are inherent to the Plan

Discussion

There is no doubt that implementing IWRM presents challenges, yet it is worth pursuing. Experiences from the Grand River watershed illustrate that an integrated approach, which considers a multi-disciplinary, multi-agency approach to concurrently address a range of resource issues, goals and outcomes is possible and can be put into practice, yielding successful outcomes. However, integration takes time, tenacity, perseverance and a commitment to continuous learning and improvement. A key tool for encouraging integration and commitment was the Project Charter and recognition of the factors for success embedded in it. The following discussion illustrates the influence of these factors on the outcomes of the WMP.

Sustaining collaboration among partners

Water management is fragmented in Ontario. Roles and responsibilities are mandated by federal and provincial legislation and are shared among many government agencies and departments, municipalities and the GRCA (Veale, 2004). Water is also central to the culture of First Nations peoples. Coordination and collaboration among and within agencies is needed to achieve an integrated approach. One of the biggest hurdles in successful watershed management is building processes that are collaborative, yet streamlined, to match stakeholder capacity and sustain interest and enthusiasm over the long term. This requires conscious effort and a commitment and willingness to pool resources and work collectively to resolve issues of mutual concern. Partners contributed to a collaborative mindset by signing the Project Charter, which clearly spells out roles, responsibilities, guiding principles and expectations. The governance structure requires ongoing communication and

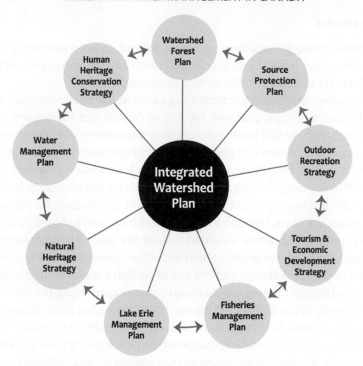

Figure 3. Components of the Integrated Grand River Watershed Management Plan.

collaboration among partners. As the secretariat, the GRCA will continue to organize meetings, collate and share information, write technical documents, and report on progress towards WMP objectives.

The preparation of the WMP began in 2010 and was endorsed in 2014; the five-year period has helped build good working relationships among the partners. There is a commitment among the WMP partners to continue to meet to discuss watershed water issues. The Water Managers Working Group collaborates through quarterly meetings. In 2015, the working group produced its first action report summarizing progress to date (Grand River Conservation Authority, 2015). In addition, there are commitments to host technical workshops to gather new knowledge and advance actions, and report on the status of implementation, which will enable partners to see progress. Evaluating resource conditions and trends at regular intervals (e.g. every five years) should provide insight into whether the partners' collective actions are achieving the goals of the WMP. The challenge will be to ensure that this assessment takes place and is communicated to all partners.

Experience demonstrates that collaboration is often the product of personal relationships, which are honed over time as an outcome of working with others to achieve common goals. Collaboration into the future will depend on the degree to which current participants are able to transfer their knowledge and experiences to their successors and the opportunities afforded to forge new relationships. As the watershed agency prescribed to facilitate agency partnerships at the watershed scale and given the aforementioned prerequisites, it is incumbent on the GRCA to continue to devote time and energy towards fostering good working relationships and establishing commitments among new participants.

Celebrating success

Sustaining interest throughout a long process is difficult. To keep partners motivated and engaged, it is important to celebrate accomplishments and share experiences to foster further action. Several examples are offered that illustrate this point.

First, the successes achieved in implementing a wastewater treatment plant optimization programme by two municipal partners, the city of Guelph and Haldimand County, were introduced to other municipal WMP partners. This programme uses the principles of the Composite Correction Program and guides wastewater operators and managers by helping them identify design, administrative or operational/maintenance barriers to achieving high-quality effluent (Hegg, DeMers, & Barber, 1996). Using this approach, both municipalities realized sufficient improvement in sewage effluent quality to defer millions of dollars in capital infrastructure costs. Through a series of workshops and hands-on training to transfer skills, a community of practice emerged around the optimization concept, and other watershed municipalities have implemented individual optimization programmes.

Second, the Region of Waterloo and the city of Guelph are established leaders in water conservation, and representatives were willing to share experiences. The GRCA hosted two workshops to identify opportunities and barriers to implementing water conservation methods and techniques, share best practices and develop a toolkit. The toolkit has been completed and includes a decision-making matrix to help municipalities identify an appropriate suite of demand management tools, based on their specific water constraints and objectives. Several 'primers' were also developed, based on shared learning among participating municipalities. These tools are being used as a basis for continuing dialogue with watershed municipalities. Several municipalities have considered and piloted the 'soft path' approach (Brooks, Maas, Brandes, & Brandes, 2015) in water supply plans, and some have set new (lower) water demand management objectives. The soft path is an alternative approach to the traditional supply-focused water planning. Rather than viewing water as an end product, the soft path considers water a means to accomplish specific tasks and changes the role of water management from merely investing in capital infrastructure to providing water-related services such as drought-resistant landscapes, low-impact development for stormwater management, and alternative ways of providing sanitation that require less water to function.

Third, collaborative programmes to tackle water quality issues associated with non-point sources of pollution have been implemented since 1998 (Project Team, Water Management Plan, 2014). The Rural Water Quality Program encourages the agricultural community to voluntarily implement best management practices (BMPs). The programme helps farmers prepare nutrient management plans and provides financial incentives for BMPs. A review of BMPs was undertaken by the WMP partners to develop effective strategies for expanding the geographic scope of the programme and encouraging uptake. Mapping tools and approaches using 3-D hydrography and high-resolution digital elevation models were developed to identify erosion source areas for remediation. The presentation of model results to the agriculture community has re-engaged many farmers, who have shown renewed interest in implementing BMPs.

Although celebrating 'early wins' is fundamental to building momentum and partner engagement, sharing challenges and barriers to implementation is equally insightful. Deliberate and focused discussion was facilitated at Project Team meetings on sensitive issues, such as municipal water supply security and the cumulative impact of multiple

wastewater discharges on water quality. Mindful, honest and open discussion allowed partners to tackle and remove barriers to improved water management in the watershed.

Scoping the plan

The Project Charter emphasized that the WMP needed to be scoped according to available
time, funding, human resources, data and science. Rather than undertaking new studies,
collecting new data or developing additional computer models, the planning process was
designed to be cost-effective, building on a wealth of existing information, available expertise and the collective knowledge of the partners. It also took into account the many programmes, activities, models and studies undertaken since the completion of the Basin Study
(GRCA, 1982).

For instance, the Grand River Simulation Model was developed in concert with the Basin
Study. It is a continuous, dynamic dissolved oxygen model that takes into account contributions from point and non-point sources of nutrients and oxygen-consuming materials to
predict in-river aquatic plant growth and associated dissolved oxygen levels (Willson, Kwong,
Weatherbe, & Post, 1982). The GRCA, in collaboration with watershed municipalities, provides
data to validate and calibrate the model. The model is used to evaluate the cumulative effects
of the 10 wastewater treatment plants that discharge effluent to the central watershed, and
to guide wastewater master planning. The previously calibrated and validated model was a
valuable predictive tool for the WMP. Modelling scenarios were run for planned and future
wastewater plant upgrades and optimization. The results showed that improvements in
water quality can be expected when all planned wastewater treatment plant upgrades are
completed. Even greater water quality improvements are predicted when the planned
upgrades to wastewater facilities are combined with plant optimization. This analysis was
instrumental in confirming the value of wastewater treatment upgrades and convincing
partners that plant optimization was a worthwhile and practical strategy.

Since the 1982 Basin Study, several watershed-scale studies have improved the understanding of hydrological processes and water quality. A regional groundwater characteriza-
tion study (Holysh, Pitcher, & Boyd, 2000) and an integrated water budget (AquaResources
Inc., 2009a) investigated and described the interactions between surface and groundwater.
The Drinking Water Source Protection Program, which is administered by the Ministry of the
Environment and Climate Change under the Clean Water Act (2006), funded a watershed
characterization report to support the drinking water source protection planning process
(Lake Erie Source Protection Region Technical Team, 2008). Water quality conditions and
trends in the watershed provided comprehensive assessments of the state of the water
resource (Cooke, 2006; Loomer & Cooke, 2011). The knowledge gained through an analysis
of these reports and others provided new insights into what actions or management options
might be appropriate for the WMP.

For example, the integrated water budget and the Water Quantity Stress Assessment
Report (AquaResources Inc., 2009b) highlighted the Norfolk Sand Plain, south-west of
Brantford, as an area of potential water stress. Agricultural irrigation, using water from
groundwater-fed creeks, is common in this area. With the support of WMP partners, Farm &
Food Care Ontario and others, a successful pilot project was initiated to move irrigation
water sources from direct in-creek withdrawals to reconditioned, abandoned off-line ponds,
reducing direct impacts on creek water levels. This pilot project effectively mobilized the

collective knowledge gained through previous technical studies, leveraged funding opportunities and built stronger working relationships and trust among the local farming community.

While an abundance of information and studies were available, the WMP partners recognized that there were some fundamental knowledge gaps. For instance, partners acknowledged the need to better understand the relationships among the physical, chemical and biological attributes of the river system. When the Canadian Water Network, a national centre of excellence that connects water research and knowledge to decision makers, issued a call for research proposals to establish a regional (watershed) monitoring framework focused on cumulative aquatic effects, the partners took advantage of the opportunity to address a key research gap and submitted a successful research proposal.

The Canadian Water Network provided CAD 600,000 to fund a three-year, multi-disciplinary research project in the watershed. The research will identify specific biotic indicators that can detect change induced by watershed stressors (e.g. population growth, land use change, changing climate) or management approaches (e.g. reservoir operations) and recommend an aquatic biological monitoring approach that builds on existing water quality and river flow monitoring. The results were received in 2015. It is anticipated that these findings will provide the cornerstone for an integrated monitoring framework for cumulative aquatic effects assessment in the Grand River watershed for consideration by the WMP partners.

Continuous improvement

The Project Charter emphasized the need for continued monitoring, assessment and reporting to support adaptive management and decision making. Adaptive management, in the context of the WMP, is an ongoing process for continually improving policies, programmes, and practices by learning from the outcomes of management actions. To determine whether management actions are having the desired effect on river health, a number of water quality indicators were selected, and associated targets and milestones for these indicators were developed (Grand River Water Management Plan, 2012, 2013).

Milestones are quantitative descriptions of future water quality conditions that are expected to result from the specific actions undertaken over a given timeframe (Project Team, Water Management Plan, 2014). Milestones will be used to gauge progress, whereas *targets* represent the end condition that supports ecosystem health. A review and rationalization of the collective monitoring undertaken by partners is required to determine whether there are sufficient data and information to report on the stated milestones.

The partners have acknowledged that the goals, objectives and targets set in the WMP may take decades to achieve. A key difference between the WMP and the Basin Study is the commitment by partners to an adaptive approach. This introduces a new generation of IWRM for the Grand River watershed. Regular reporting to assess the status of plan actions and on-the-ground effects is critical to continued engagement and a key component of the plan. Each partner has committed to providing regular updates to the Water Managers Working Group on the progress of their actions and any additional activities they have initiated in support of the WMP.

Currently, there is a commitment that the GRCA will assist the Water Managers Working Group in preparing an annual implementation progress report, starting in 2015, and assembling a technical watershed report on the progress towards meeting milestones and targets every five years. These assessments will be used to adjust implementation strategies and actions to make them more effective and economical. Further, as knowledge gaps are addressed and new data, science, technologies or approaches are developed, there will be opportunities to adjust and improve the plan.

Conclusions

Many agencies in Ontario share responsibility for implementing IWRM. Collaboration among decision makers is required to align actions to achieve common goals. While innovative technologies and infrastructure investments can help solve water problems, the people responsible for all aspects of water management (the implementers) are central to achieving the goals. A coordinating agency, which can oversee the planning process, engage and galvanize partners, ensure effective use of time and resources and manage partner expectations, is critical to the success of a collaborative approach.

The process of updating the WMP brought the implementers to the table, fostered good working relationships and elicited respect and trust among the partners. This provided a foundation for open communication and shared experiences. The success of the process was manifested in well-attended workshops and meetings, synthesis of existing technical studies, update of computer models, development of communities of practice, commitment by the partners to undertake actions, and a formal endorsement of the final WMP.

The WMP was voluntary, driven by the desire of partners to continually improve and strive for common goals, in the public interest. A voluntary, collaborative process can be more progressive and innovative than a regulatory approach, since there is little risk to implementers in setting the bar just a bit higher and little fear in making adjustments if things change. Furthermore, the implementers were willing to align their actions with others to achieve the goals of the WMP because they maintained control of activities that affected them. In addition, actions were discussed in the context of 'best value' solutions – those actions that will give the best return on investment (although it was acknowledged that a more robust economic analysis is required). Many partners were willing to try new management approaches, such as the Composite Correction Program, based on the experiences and recommendation of other partners, and were able to realize cost savings through improved operations or deferral of capital infrastructure investments.

The Project Charter was the primary mechanism for obtaining initial buy-in and was a crucial first step towards collaboration by setting out the collective goals and guiding principles for partners. While the influence of several individuals who provided important support and leadership by sharing information with their agency counterparts and securing financial support is essential, the success of the WMP thus far is largely due to the commitment of the GRCA to facilitate and administer the process and rally the partners.

Sustaining a collaborative approach into the future will continue to be a challenge as agencies evolve and partner representatives change, and as fiscal, social and environmental priorities unfold. Regarding the former, a commitment to succession planning is needed so that knowledge is captured and transferred to new participants. Lack of such planning was one of the failings of the Basin Study. Long-term institutional commitment from all partners is required and needs to be continually fostered at all levels within partner agencies (e.g. technical staff and senior managers).

It is hoped that by taking an adaptive management approach, the value of voluntary collaboration will be reinforced through the process of monitoring, assessing and reporting on progress and successes. The commitment of the partners to continuous improvement and regular plan review and renewal is how the goals and targets for the Grand River watershed will ultimately be achieved. The challenge lies in maintaining the engagement of the partners and the commitment of the GRCA to continue to support the Water Managers Working Group despite competing priorities and periods of fiscal austerity.

Integrated watershed management is not easy, but it is a concept that has a multitude of practical applications that can achieve positive environmental, social and economic results, as illustrated by the successes achieved in the Grand River watershed.

Notes

1. Under the federal Indian Act, an 'Indian reserve' is land held by the Crown 'for the use and benefit of the respective bands for which they were set apart' (Section 18 (1)). Two reserves, the Six Nations of the Grand River Territory and the Mississaugas of the New Credit, are located just south of Brantford. The Six Nations reserve is the largest in Canada, with a land base of 18,000 ha and a population of over 25,000 people.
2. In 1973, the towns of Galt, Preston and Hespeler amalgamated to become the city of Cambridge.
3. The Mississaugas of the New Credit reserve does not obtain drinking water from sources in the Grand River watershed; rather, it is served by water drawn from Lake Erie, treated and distributed by pipe from Nanticoke, Ontario (Dupont et al., 2014).
4. Ontario has three types of municipalities, which include upper- and lower-tier municipalities within a two-tier structure, and single-tier municipalities. In the Grand River watershed, there are 7 upper-tier municipalities, 27 lower-tier municipalities and 5 single-tier municipalities.
5. Resilience, in the context of the WMP, refers to the longer-term capacity of a natural system or watershed to deal with change, either gradual or sudden, such as a large storm event, and continue to function as expected. Increasing the resiliency of a watershed requires new or modified beneficial practices, the safeguarding of green infrastructure, and improved management approaches to maintain or enhance the watershed's natural ability to function as expected (Project Team, Water Management Plan, 2014).
6. A Lakewide Action and Management Plan is a plan of action to assess, restore, protect and monitor the ecosystem health of a Great Lake. It is used to coordinate the work of all collaborating partners to improve the lake ecosystem.

Acknowledgements

The authors wish to acknowledge the leadership provided by L. Minshall for directing the water management plan. The water management plan could not have been completed without the contributions and support from the following staff: D. Boyd, J. Etienne, M. Anderson, C. Holeton, S. Shifflett, A. Wong, B. McIntosh, Z. Green, J. Marshall, A. Loeffler, L. Heymig, T. Ryan, D. Schultz, K. Balpataky, H. Kovacs, E. Fanning, S. Strynatka, J. Pitcher and J. Farwell.

Disclosure statement

No potential conflict of interest was reported by the authors.

Funding

This work was supported by the Showcasing Water Innovation programme of the Ontario Ministry of Environment and Climate Change [grant number SWI-11-12-0119].

References

Adams, F. (1937). Water supply and sewage disposal to be aided by flood control measures on the Grand River. *Engineering and Contract Record, 50*, 19–22.

AquaResources Inc. (2009a). *Integrated water budget*. Final version June 2009. Grand River Conservation Authority. Retrieved February 13, 2015, from http://www.sourcewater.ca/swp_watersheds_grand/Grand_2009WaterBudget_final.pdf

AquaResources Inc. (2009b). *Tier 2 water quality stress assessment report*. Final version December, 2009. Grand River Conservation Authority. Retrieved February 9, 2015, from http://www.sourcewater.ca/swp_watersheds_grand/Grand_2009Stress_Final.pdf

Biswas, A. K. (2004). Integrated water resources management: A reassessment. *Water International, 29*, 248–256.

Biswas, A. K. (2008). Integrated water resources management: Is it working? *International Journal of Water Resources Development, 24*, 5–22.

Blomquist, W., & Schlager, E. (2005). Political pitfalls of integrated watershed management. *Society and Natural Resources, 18*, 101–117.

Brooks, D. B., Maas, C., Brandes, O. M., & Brandes, L. (2015). Applying water soft path analysis in small urban areas: Four canadian case studies. *International Journal of Water Resources Development, 31*, 750–764. doi: 10.1080/07900627.2014.995265.

Butterworth, J., Warner, J., Moriarity, P., Smits, S., & Batchelor, C. (2010). Finding practical approaches to integrated water resources management. *Water Alternatives, 3*, 68–81.

Clean Water Act. (2006). Retrieved February 16, 2015, from http://www.e-laws.gov.on.ca/html/statutes/english/elaws_statutes_06c22_e.htm

Conservation Authorities Act. (1990). Retrieved February 15, 2015, from https://www.e-laws.gov.on.ca/html/statutes/english/elaws_statutes_90c27_e.htm

Cooke, S. (2006). *Water quality in the Grand River: A summary of current conditions (2000–2004) and long-term trends*. Grand River Conservation Authority. Retrieved February 13, 2015, from http://www.grandriver.ca/water/2006_WaterQuality_complete.pdf

Cortner, H., & Moote, M. (2000). *Ensuring the common for the goose: implementing effective watershed policies*. USDA Forest Service Proceedings RMRS–P–13, 247–256. Retrieved February 2, 2015, from http://www.fs.fed.us/rm/pubs/rmrs_p013/rmrs_p013_247_256.pdf

Dupont, D., Waldner, C., Bharadwaj, L., Plummer, R., Carter, B., Cave, K., & Zagozewski, R. (2014). Drinking water management: Health risk perceptions and choices in first nations and non-first nations communities in canada. *International Journal of Environmental Research and Public Health, 11*, 5889–5903.

Global Water Partnership Technical Advisory Committee. (2000). *Integrated water resources management*. TAC Background Papers No. 4. Stockholm: Global Water Partnership.

Goldstein, J., & Huber-Lee, A. (2004). *Global lessons for watershed management in the United States*. Boston, MA: Tellus Institute.

Grand River Conservation Authority (GRCA). (1979). *Environmental assessment of water control structures in the Grand River basin*. Prepared by the Steering Committee for the Grand River Conservation Authority. Cambridge: Grand River Conservation Authority.

Grand River Conservation Authority (GRCA). (1994). *The grand strategy for managing the Grand River as a Canadian Heritage River*. Cambridge: Grand River Conservation Authority.

Grand River Conservation Authority (GRCA). (1997). *Focus on watershed issues 1996–1997: State of the Grand River watershed*. Cambridge: Grand River Conservation Authority.

Grand River Conservation Authority (GRCA). (2009). National fish award. *Grand Actions, 14*(3), 3.

Grand River Conservation Authority (GRCA). (2010). *Grand River watershed water management plan project charter*. Cambridge: Grand River Conservation Authority.

Grand River Conservation Authority (GRCA). (2014). *Grand river water management plan*. Retrieved from http://www.grandriver.ca/index/document.cfm?Sec=87&Sub1=0&sub2=0

Grand River Conservation Authority (GRCA). (2015). Grand river watershed water management action plan: 2014 report on actions. Retrieved November 23, 2015, from http://www.grandriver.ca/waterplan/2014_ReportOnActions.pdf

Grand River Disaster Relief Committee. (1975). *Report, June 11*. Cambridge: Grand River Conservation Authority.

Grand River Fisheries Management Plan Implementation Committee. (2005). *Grand River fisheries management plan*. Cambridge: Grand River Conservation Authority.

Grand River Implementation Committee. (1982). Grand River basin water management study. Retrieved February 2, 2015, from http://www.grandriver.ca/WaterPlan/1982_BasinStudy.pdf

Grand River Water Management Plan. (2012). *A framework for identifying indicators of water resource conditions: Support of ecological health by water resources in the Grand River-Lake Erie interface*. Prepared by the Water Quality Working Group, Grand River Conservation Authority. Retrieved February 16, 2015, from http://www.grandriver.ca/waterplan/WaterResourceIndicatorFramework_V2.pdf

Grand River Water Management Plan. (2013). *Water quality targets to support healthy and resilient aquatic ecosystems in the Grand River watershed*. Prepared by the Water Quality Working Group. Grand River Conservation Authority, Retrieved February 16, 2015, from http://www.grandriver.ca/waterplan/waterQualityTargetsFeb192013.pdf

Heathcote, I. (2009). *Integrated watershed management: Principles and practice* (2nd ed.). Hoboken, NJ: John Wiley & Sons Inc.

Hegg, B., DeMers, L., & Barber, J. (1996). *The Ontario composite correction program manual for optimization of sewage treatment plants*. Draft report, prepared for the Ontario Ministry of Environment and Energy, Environment Canada and The Municipal Engineers Association. Unpublished.

Holysh, S., Pitcher, J., & Boyd, D. (2000). *Regional groundwater mapping: An assessment tool for incorporating groundwater into the planning process*. Grand River Conservation Authority, Retrieved February 13, 2015, from http://www.grandriver.ca/groundwater/cwra_paper.pdf

Indian Act. (1985). In *Revised Statutes of Canada*, 1985. Retrieved May 15, 2015, from http://laws-lois.justice.gc.ca/eng/acts/i-5/page-10.html#h-13

Krause, P., Smith, A., Veale, B., & Murray, M. (2001). Achievements of the grand river conservation authority, Ontario, Canada. *Water Science and Technology, 43*, 45–55.

Lake Erie Source Protection Region, Technical Team. (2008). *Grand River Watershed characterization report. grand river conservation authority*. Retrieved February 20, 2015, from http://www.sourcewater.ca/swp_watersheds_grand/Characterization_Grand.pdf

Leach, W. W. (1975). *Report of the Royal Commission inquiry into the Grand River Flood, 1974*. Toronto: Queen's Printer.

Loomer, H., & Cooke, S. (2011). Water quality in the Grand River watershed: Current conditions & trends (2003 – 2008). Grand River Conservation Authority. Retrieved February 13, 2015, from http://www.grandriver.ca/water/2011_WaterQualityReport.pdf

Mitchell, B. (2009). Implementation Gap. *IWRA Update, 22*, 7–12.

Mitchell, B. (2015). Water risk management, governance, IWRM and implementation. In F. Urbano (Ed.), *Risk governance: The articulation of hazard, politics and ecology* (pp. 317–335). Netherlands: Springer.

Mitchell, B., & Shrubsole, D. (1992). *Ontario conservation authorities: Myth and reality*. Waterloo: University of Waterloo, Department of Geography Publication Series No. 35.

Molle, F. (2008). Nirvana concepts, narratives and policy models: Insights from the water sector. *Water Alternatives, 1*, 131–156.

Nelson, J. G., & O'Neill, P. C. (1989). *The grand as a Canadian heritage river occasional paper 9*. Waterloo: University of Waterloo, Heritage Resources Centre.

Ontario Department of Lands and Forests. (1932). *Report on Grand River drainage*. Toronto: King's Printer.

Ontario Department of Lands and Forests. (1962). *Grand River conservation report - hydraulics* (2nd ed.). Toronto: Ontario Department of Lands and Forests, Conservation Authorities Branch.

Ontario Sewer and Watermain Construction Association. (2001). *Drinking water management in Ontario: A brief history*. Retrieved February 10, 2015, from http://www.archives.gov.on.ca/en/e_records/walkerton/part2info/publicsubmissions/pdf/drinkingwaterhistorynew.pdf

Ontario Treasury Board. (1971). *Review of planning for the Grand River watershed Project #229*. Toronto: Ontario Treasury Board.

Plummer, R., Spiers, A., FitzGibbon, J., & Imhof, J. (2005). The expanding institutional context for water resources management: The case of the grand river watershed. *Canadian Water Resources Journal, 30*, 227–244.

Project Team, Water Management Plan. (2014). Grand River watershed water management plan. Retrieved February 2, 2015, from http://www.grandriver.ca/waterplan/2014_WMP_Final.pdf

Saravanan, V., McDonald, G., & Mollinga, P. (2009). Critical review of integrated water resources management: Moving beyond polarised discourse. *Natural Resources Forum, 33*, 76–86.

Shaver, E., Horner, R., Skupien, J., May, C., & Ridley, G. (2007). *Fundamentals of urban runoff management: Technical and institutional issues*. Madison, WI: North American Lake Management Society.

Veale, B. (2004). Watershed management in the Grand River watershed. In J. G. Nelson, B. Veale, & B. Dempster (Eds.), *Towards a Grand Sense of Place* (pp. 261–267). Waterloo: University of Waterloo, Department of Geography.

Veale, B. (2010). *Assessing the influence and effectiveness of watershed report cards on watershed management: A study of watershed organizations in Canada*. (unpublished doctoral dissertation). Waterloo: University of Waterloo, Department of Geography.

Veale, B., Cooke, S., Zwiers, G., & Neumann, M. (2014). The Waterloo Moraine: A watershed perspective. *Canadian Water Resources Journal, 39*, 181–192.

Water Opportunities Act. (2010). Retrieved from February 15, 2015, http://www.e-laws.gov.on.ca/html/statutes/english/elaws_statutes_10w19_e.htm

Water and Sewage Services Improvement Act. (1997). Retrieved February 15, 2015, from http://www.ontla.on.ca/web/bills/bills_detail.do?locale=en&BillID=1493&ParlSessionID=36:1&isCurrent=false

Willson, K., Kwong, A., Weatherbe, D. G., & Post, L. (1982). *Water quality simulation models and modelling strategy for the Grand River basin*, Grand River Basin Water Management Study Technical Report # 30, April, 1982 (edited and published June, 1996). Toronto: Ontario Ministry of the Environment, Water Resources Branch.

Lessons from implementing integrated water resource management: a case study of the North Bay-Mattawa Conservation Authority, Ontario

Paula Scott, Brian Tayler and Dan Walters

ABSTRACT

This case study explores the North Bay-Mattawa Conservation Authority's experience in implementing IWRM. Successes include protecting life and property by mitigating flood and erosion hazards; building capacity through multi-stakeholder collaborations; and fostering community stewardship. Ongoing challenges include limited resources and narrow mandate for addressing broader watershed and natural resources issues; and a need to enhance relationships with First Nations. The NBMCA has learned numerous lessons on how to apply IWRM, including collaborating early and often and fostering community stewardship.

Introduction

This article has two objectives. First, this case study explores the experience of the North Bay-Mattawa Conservation Authority (NBMCA) in implementing integrated water resource management (IWRM). This case study contributes to the literature investigating aspects of Ontario's conservation authority programme (Bullock & Watlet, 2006; Mitchell, Priddle, Shrubsole, Veale, & Walters, 2014; Mitchell & Shrubsole, 1992; Shrubsole, 1990, 1996). In Ontario, conservation authorities are watershed-based organizations created under the Conservation Authorities Act (1946), which establishes a partnership between the province and local member municipalities. There are currently 36 conservation authorities in Ontario. The NBMCA is one of five in northern Ontario. Each conservation authority is unique by virtue of developing independently and under different local circumstances (NBMCA, 1982b). Additional roles have been delegated to conservation authorities through other legislation, such as the Ontario Building Code Act, S.O. (1992) or Clean Water Act, S.O. (2006). Following other studies on conservation authorities (Bullock & Watlet, 2006; Mitchell & Shrubsole, 1992; Mitchell et al., 2014; Shrubsole, 1996), we use some of the founding principles to frame our interpretation of the successes and challenges: (1) local initiatives and involvement; (2) municipal–provincial partnership; and (3) integrated and coordinated management on a watershed basis (NBMCA, 1983b). These three principles, explicitly and implicitly, embody the principles of IWRM: watershed units, surface–groundwater interaction, connections with other natural resources, and multi-stakeholder collaboration (Mitchell et al., 2014).

Second, this case study contributes to the discussion and debate concerning the IWRM paradigm (Pahl-Wostl, Jeffrey, Isendahl, & Brugnach, 2011). This paradigm requires a fundamental transformation of water governance structures, with key agents and networks providing leadership and vision for new institutional alternatives (Pahl-Wostl, 2009). Yet, successful implementation of IWRM has often proven difficult. This has led some researchers to question the effort and expense of transitioning to an integrated watershed approach when the empirical evidence is inconclusive (Biswas, 2008; Giordano & Shah, 2014; Grigg, 2014). However, others (Mitchell, 1990; Lawrence, 2011) argue that the principles of IWRM are not the issue; the limited resources, institutional arrangements, commitment and social capital of watershed organizations hamper implementation, and exist independently of IWRM (Patterson & Smith, 2013). The conservation authority programme is a real-world situation used here to examine IWRM. Implementation barriers are discussed in the context of how IWRM is taking place in the NBMCA. Several of the NBMCA's watershed planning initiatives are examined to illustrate the lessons learned. We do not begin with an assumption that IWRM is the ideal paradigm; the principles of IWRM frame the context within which the conservation authorities operate.

The article is in four sections. The first considers how strategic initiatives and institutional changes in Ontario since 1972 have influenced conservation authorities' capacity to implement IWRM. The second describes the NBMCA case study and uses some of the founding principles to frame a discussion of the successes and challenges in implementing IWRM. The third describes the lessons learned about IWRM by comparing and contrasting watershed planning exercises by the NBMCA. The fourth concludes by identifying strategies for addressing IWRM implementation barriers.

The article is written from the practitioners' perspective, and thus we acknowledge that it is difficult to separate the researchers from the research. Our interpretations are based on a critical reflection on perceived successes and challenges of the NBMCA. The archives at the NBMCA are the primary source of materials used to form our interpretation of the NBMCA experience. The sources of data include internal documents, reports, plans, public speeches and other archived materials. Some internal documents include statements by board members and the general public. However, the majority of insights are attributed to staff at the NBMCA who contributed to the critical reflections.

Strategic initiatives and structural adjustments since 1972

Mitchell and Shrubsole (1992) provide details on the origins of the 1946 Conservation Authorities Act in Ontario. The act represented a major initiative that signalled a new approach in Ontario to conservation and resource management that embodied elements of IWRM principles, either explicitly or implicitly: local initiatives and involvement; municipal–provincial partnership; and integrated and coordinated planning at the watershed scale. The act was the mechanism by which municipalities could work together to solve their resource development conflicts. Municipal concerns could be identified and tackled at a local level, with financial assistance from the province. The conservation authority represents a partnership among member municipalities, the province and the public. The municipal representatives and public voice the interests of local residents, while the province ensures that programmes and projects conform to provincial policies (NBMCA, 1982a). Public interests are sought through membership on advisory boards or other public sessions.

The mandate and regulatory powers of the Conservation Authorities Act (1990) are very broad. Section 20(1) stipulates that the objects of an authority are "to establish and undertake, in the area over which it has jurisdiction, a program designed to further the conservation, restoration, development and management of natural resources other than gas, oil, coal and minerals" (R.S.O. 1990, c. C.27, s. 20). The scope of the regulatory powers of the conservation authority is outlined in Section 28(1): (a) regulating the use of water; (b) regulating the straightening, changing, diverting or interfering in any way with a watercourse; (c) regulating development if deemed to impact flooding, erosion, dynamic beaches, pollution or conservation lands (1998, c. 18, Sched. I, s. 12). In addition, conservation authorities, in partnership with member municipalities and the province, can assume lead agency roles in water management and conservation of natural resources within their jurisdictional watersheds. For example, Section 21(1)(a) empowers the conservation authority "to study and investigate the watershed and to determine a program whereby the natural resources of the watershed may be conserved, restored, developed and managed". Also, conservation authorities can be delegated responsibility to represent provincial interests by acting as a regional public commenting body on applications under the Planning Act, R.S.O. (1990), Environmental Assessment Act, R.S.O. (1990), and others. The flexibility and possible range of jurisdictional responsibility in the Conservation Authorities Act, R.S.O. (1990), has meant that conservation authorities have developed largely independently of one another. Since the 1946 Conservation Authorities Act, the conservation authorities also have been given other roles, especially since the Walkerton Commission of Inquiry (discussed later). Some of these changing responsibilities are discussed in the case study section. However, strategic programme initiatives and institutional changes have influenced all conservation authorities' capacity to implement IWRM.

We focus here on the programme and institutional changes since the formation of the NBMCA in 1972. First, the jointly sponsored (federal and provincial) Flood Damage Reduction Program (FDRP), which started in 1975, funded floodplain and fill-line mapping across Canada. The aim of the FDRP was to reduce property damage and loss of life along the shores of rivers and lakes by mapping flood-risk areas and discouraging development. This programme certainly influenced the activities of the conservation authorities from 1975 to the early 1990s. As flood control has always been a core responsibility of conservation authorities, they took the lead in coordinating the mapping of hazard lands. The FDRP ended in 1998. The funding helped reduce the risk of flooding in populated areas in Ontario, and had administrative, social and environmental benefits (Millerd, Dufournaud, & Schaefer, 1994). FDRP funding helped build capacity with respect to protecting life and property from the risk of flooding and erosion. The flood and erosion control sub-watershed studies also generated useful baseline information for other watershed management efforts.

Second, structural adjustments and funding cuts by the provincial government between 1987 and 1997 created a turbulent period for conservation authorities. In 1987, the provincial government published a *Review of the Conservation Authorities Program* that reported that nearly a third of conservation authorities were receiving almost 80% of their annual revenue from the provincial government. In 1991, the Ontario Ministry of Natural Resources distinguished core and non-core programming to narrow the responsibilities and funding of conservation authorities. The core programming included flood and erosion control, conservation areas and conservation information. The non-core activities included outdoor education and programme awareness. Also, the cost-sharing adjustments included phasing out provincial supplemental grants. Provincial funding of conservation authority programmes across the province fell from CAD 33 million in 1995 to CAD 12 million in 1996, and CAD 8 million by 2002 (Mitchell et al., 2014). The funding adjustments meant that each conservation authority had to seek alternative means to support core operations. Whereas the FRDP helped build capacity, the structural adjustments narrowed the scope of conservation authorities' mandate and significantly reduced financial support. This restricted the capacity of conservation authorities to implement IWRM. We discuss below the NBMCA experience in adapting to the new provincial funding arrangement.

Here the NBMCA is used to illustrate how the structural changes influenced the funding sources. Initially, the municipalities were motivated to form a conservation authority to receive new funding from the province. The NBMCA member municipalities could leverage 3 to 4 times the provincial funding relative to their contributions. However, the nearly 75% provincial funding cut in the mid-1990s left conservation authorities asking how to finance basic day-to-day operations. In 1986, the total annual revenue for the NBMCA was CAD 874,621, with provincial funds at 67%, municipal levy 14%, other grants (federal/provincial) 18%, and self-generated revenue less than 1%. In 2013, total annual revenue was approximately CAD 3 million, with provincial funds at 26%, municipal levy 33%, other grants (federal/provincial) less than 1%, self-generated revenues 22%, and carry-over surplus 17%. Municipal levy and self-generated revenues together now make up the greater percentage of revenues. The total municipal levy in 1986 was approximately CAD 120,000, whereas in 2013 it was CAD 990,000. Technical service agreements with municipalities and provincial ministries helped offset some of the funding cuts. To cover the operating costs of the organization, 'fees for services' became a part of the NBMCA operations through fill regulations and septic inspections. Furthermore, in 1993, the NBMCA received official status as a charitable organization. This enables the authority to issue tax-deductible receipts to individuals and businesses making donations.

Not all the alternative funding strategies have been as successful as hoped. For example, the Jack Pine Hill Multi-Use Four Season tourism initiative has not been financially successful. The aim was to extend beyond the winter operations – skiing, snowboarding and snow tubing – by building infrastructure to accommodate hikers, cyclists and nature lovers in the spring, summer and fall. The initiative was to include an interpretive centre to demonstrate the area's geological and natural history. Jack Pine Hill, now known as the Laurentian Ski Hill, is owned by the NBMCA, but the ski operation is run by a separate business. The annual operating costs of the Laurentian Ski Hill are approximately CAD 600,000. The NBMCA helps cover some administrative costs (e.g. phone, internet, heating) and some capital improvements to the ski hill. For the most part, however, the ski hill venture has been a success. The community has a natural environment conservation area and a ski hill in the city of North Bay. Outside the winter season, the area is used for conservation and education programmes. These would not have been possible without the investment in the facility.

The third structural adjustment occurred in response to the Walkerton Commission of Inquiry recommendations (O'Connor, 2002a, 2002b). Chief Justice O'Connor recommended that conservation authorities be responsible for coordinating source water protection committees and developing watershed-based drinking water protection plans (Mitchell et al., 2014). Under the Clean Water Act, conservation authorities coordinate the development of source water protection plans for municipal drinking water systems. Tasks include coordinating preparation of technical reports, compiling technical and scientific data, facilitating local engagement and communications, and allocating funding. The Ontario provincial government invested CAD 120 million between 2004 and 2008 in the conservation authority budget to establish source water protection plans, and another CAD 21 million to support education, outreach, and incentives by the source water protection committees to protect municipal sources of drinking water (Mitchell et al., 2014). The financial investment helped rebuild some of the financial capacity the conservation authorities had lost in the 1990s. However, this type of special initiative funding is often short-term and does not help with long-term implementation efforts. For example, the Clean Water Act is narrowly scoped to address threats to sources of municipal drinking water supplies in Ontario.

These three phases affected how conservation authorities were applying IWRM. The FDRP and Drinking Water Source Protection Program (of the Clean Water Act) were special initiatives that provided financial support to help build human, technical and information capacity. Often, the commissioned studies and public engagement sessions that support these initiatives indirectly contribute to other watershed management activities and public relationship building. The conservation authorities must seek alternative support for broad watershed management functions. The structural and financial adjustments of the 1990s strained the capacity of conservation authorities to deliver core programmes. Within the context of these changes, we now focus on how the NBMCA applies the principles of IWRM.

NBMCA case study

The NBMCA was formed in 1972 under Ontario's Conservation Authorities Act (1946). It included nine municipalities in the Mattawa River watershed and the eastern portion of Lake Nipissing. The municipal representatives voted 7 to 3 in favour of forming the NBMCA. The city of North Bay had two representatives, and each of the smaller municipalities had one representative. Even though the vote was not unanimous, all nine municipalities became members of the NBMCA. Each municipality makes a financial contribution to the conservation authority. However, being part of a conservation authority, each municipality also is eligible for funding from the province, funding not accessible to a municipality operating on its own.

The NBMCA's administrative boundary is unique among Ontario conservation authorities as it straddles the headwaters of two major basins, the Ottawa River and the Great Lakes (Figure 1). The NBMCA covers 2890 km². The majority of the NBMCA's core administrative area (2295 km²) covers all sub-watersheds of the Mattawa River that flow into the Ottawa River. However, its core administrative area (595 km²) also includes all rivers between Duchesnay Creek and Bear Creek draining into Lake Nipissing and the Great Lakes watershed. The NBMCA core administrative area in the Lake Nipissing watershed has expanded on two occasions since 1972. In 1982, it expanded to include the Wistiwasing (Wasi) River watershed. In 2003, the municipality of Callander became part of the NBMCA administrative area, which now included three additional rivers. Yet, the NBMCA's jurisdiction in Lake Nipissing is less than 5% of the 12,000 km² watershed. The NBMCA is therefore limited in its capacity to lead initiatives in other threatened embayments in Lake Nipissing, such as the West Arm and Cache Bay in the western end of the lake. The NBMCA does not anticipate an expansion of its jurisdiction to the other watersheds contributing to Lake Nipissing. Since a portion of the provincial funding formula is based on municipal taxes, it would be a challenge to serve

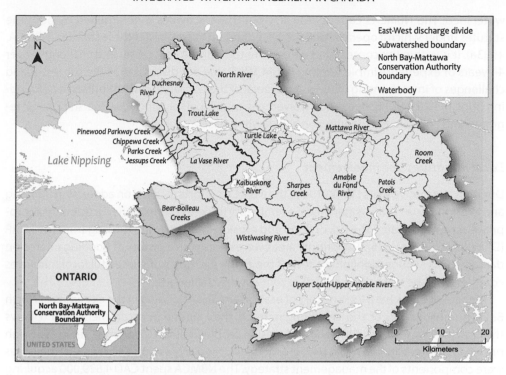

Figure 1. North Bay-Mattawa Conservation Authority Jurisdiction.

these other regions. Also, there has been no expression of interest from the other munici-palities about forming or expanding the NBMCA jurisdiction. However, this does not preclude the formation of partnerships with other entities to address the gaps in data collection, management, monitoring and reporting in other parts of the Lake Nipissing watershed. For example, Nipissing University (2016) is helping coordinate a State of the Basin Report on Lake Nipissing in partnership with the NBMCA, the Ministry of Environment and Climate Change, the Ministry of Natural Resources and Forestry, and municipal, First Nations and community groups, which may lead to further collaborations in the future.

Operationally, the NBMCA has 22 full-time employees allocated among six departments: administration; drinking water source protection; watershed planning and regulations; on-site sewage; communications and outreach; and conservation lands and property man-agement. These departments align with the three main functions of the NBMCA. First, the core duties of the NBMCA include the activities outlined in the Conservation Authorities Act (1990) discussed above (e.g. regulation of revelopment, interference with wetlands, and alterations to shorelines and watercourses). Second, in the early 1990s the NBMCA began administering Ontario Building Code (1990) inspection and sewage system permits. The NBMCA reviews on-site sewage system applications to ensure compliance with the code. The NBMCA administers this programme for several areas in the region: Nipissing and Parry Sound Districts, including Algonquin Park. The administrative area for this function is based on political boundaries, because local municipalities implement the code. Third, in 2008 the responsibility of the NBMCA expanded to include coordinating the preparation and imple-mentation of Clean Water Act source protection plans for member municipalities, plus several in the Parry Sound district (e.g., Trout Creek, South River), south of the Lake Nipissing watershed. The NBMCA's capacity to carry out these activities is determined by both the statutes under which it operates and the system set up to administer these statutes (NBMCA, 1983a). It is within this context that we explore the NBMCA's successes and challenges over 44 years of experience implementing IWRM. We frame our discussion of the successes and challenges of implementing IWRM using several of the Conservation Authorities Act's founding principles: local initiatives and involvement; municipal–provincial partnership; and integrated and coordinated planning on a watershed basis.

Local initiatives and involvement

During its formative years (1972–1986), the NBMCA's efforts focused on addressing flooding and erosion issues in the watershed. Several reports were commissioned to identify appropriate actions to reduce the risk of flood and erosion damage using funding from the FDRP (Dillon Consulting, 1975). The priority areas were rivers with a history of flooding or erosion damage. This section highlights a few of the many examples of successful local initiatives completed in the watershed, as well as some of the challenges.

Chippewa Creek has a long history of erosion and flash flooding in the city limits of North Bay. The headwaters of the creek originate on the Laurentian escarpment, and it flows through the city. Dealing with flood and erosion problems along Chippewa Creek was a high priority to protect life and avoid property damage. Land acquisition and remedial measures were components of the management strategy. The NBMCA spent CAD 4,829,000 acquiring 83 dwellings at risk of flooding (within the 25-year floodplain) and erosion hazards. Between 1978 and 1985, the NBMCA coordinated efforts with the city of North Bay to reduce flooding and erosion damage, and cooperated with the Ministry of Environment and Climate Change to address water quality issues. The province, through the Ministry of Municipal Affairs and Housing and the Ministry of Natural Resources and Forestry, selected the Chippewa Creek watershed management study as one of seven throughout the province to be a pilot project for the provincial watershed management guidelines. Each watershed study followed the provincial guidelines, while adding its own local interpretations, and provided feedback to the province on what worked and what did not. This was a commendable effort toward strengthening communication between the provincial and local agencies (W. Beckett, personal communication, 22 April 1992). Overall, the efforts to control flooding and erosion in Chippewa Creek also improved water quality. The Provincial Water Quality Monitoring Network station in Chippewa Creek shows that overall water quality is improving and dissolved oxygen levels are high; however, road salt and high water temperature in the lower reaches (exceeding 20 °C in the summer) remain a problem (NBMCA, 2013).

Flooding along Lake Nipissing's shoreline presents a risk to waterfront development. In view of the high recreational potential of the North Bay waterfront, a special shoreline management programme was implemented. Land acquisition and flood-proofing, together with breakwater protection, were implemented along the North Bay waterfront to combat flooding, wave and erosion damage. The dams controlling lake levels are overseen by a Watershed Management Advisory Group, which includes Public Works and Government Services Canada, Ministry of Natural Resources and Forestry, NBMCA, First Nations and other stakeholders. Lake Nipissing is drawn down during the winter to help reduce the risk of flooding during the spring freshet. Public Works and Government Services Canada operates three dams on the French River, the outlet of Lake Nipissing, to control the lake levels. The NBMCA monitors the flows of some of the rivers contributing to the lake, as well as snow accumulation. Water quality was another concern, as the stormwater from rivers was contributing bacteria to public beaches along the Lake Nipissing shoreline that were above the provincial guidelines for recreational water quality. Efforts to improve water quality by the city of North Bay included finding and preventing sanitary discharges into storm sewers. However, despite this effort, swimming is discouraged at some public beaches for a 24-hour period after rainfall events (NBMCA, 2013).

In addition to these projects, the NBMCA has developed a local flood forecasting / flood warning system (NBMCA, 2016b) on streams with significant potential for loss of life and property damage. The NBMCA maintains a system to monitor streamflows, rainfall, water content in snow, temperature, and soil moisture and evaporation, to determine high-risk periods for flooding. Keeping the public and news media aware and knowledgeable of the roles and activities of the NBMCA is second only to flood warning.

Many other examples exist of successful local initiatives. The NBMCA has collaborated with each member municipality to address local flooding and erosion hazards. However, because of the low population density in the administrative area, there are sub-watersheds in the NBMCA jurisdiction with limited information regarding flooding. The next section discusses this ongoing challenge in more detail.

Another remaining challenge is to foster better relations with First Nations in the region. The NBMCA administrative area lies within the traditional territories of at least five First Nations. While there has been some collaboration on initiatives, such as the Chippewa Creek EcoPath (2016), the La Vase Portage celebrations and the recent Integrated Watershed Management Strategy, involvement has been intermittent. First Nations are invited to participate in policy and planning initiatives. We cannot comment on why First Nations have not participated in previous initiatives. Elsewhere, Rizvi, Adamowski, and Patrick (2013) found that First Nations with limited financial resources had less capacity to support actor network involvement. It is evident that the NBMCA must make further attempts to establish relationships with First Nations whose traditional and treaty territories overlap the Mattawa River and Lake Nipissing watersheds. Relationships with First Nations are moving in a positive direction. However, there was limited participation in the Integrated Watershed Management Strategy (NBMCA, 2015a) planning process, which is discussed in the lessons learned section.

The provincial funding helped bring municipalities and conservation authorities together to address local resource issues and encourage local involvement. Much evidence shows that with sufficient financial support, local initiatives can be successfully implemented. However, the benefits of these initiatives to broader watershed management initiatives are largely unknown, because the NBMCA has limited capacity to monitor water conditions throughout its jurisdiction. The often narrow focus of municipal and provincial interests does present a challenge in tackling broader watershed and natural resource issues.

Municipal–provincial partnership and IWRM

Conservation authorities operate in an evolving municipal–provincial partnership rooted in the management of water quantity and water-related natural hazards (NBMCA, 2015c). The NBMCA is a partnership, now with local representatives from 11 municipalities and the province working together to further the conservation, restoration, development and management of our renewable natural resources. Section 21.(1)(n) of the Conservation Authorities Act (1990) authorizes conservation authorities to make agreements: "to collaborate and enter into agreements with ministries and agencies of government, municipal councils and local boards and other organizations". Thus, conservation authorities, in partnership with member municipalities and the province, can assume lead agency roles in water management and the conservation of natural resources in their administrative areas. However, the provincial–municipal partnership is not always the easiest to function in. "The idea of blending local and provincial priorities may sound wonderful in theory, but implementation will require careful negotiation with all participating partners" (NBMCA, 1983a, p. 92). The focus on local and provincial interests can sometimes lead to narrowly conceived, single-purpose programmes or projects. As examples, the Callander Bay / South Shore sub-watershed is too narrowly focused around blue-green algae blooms because of provincial funding for protection of drinking water through source protection plans (SPPs) (NBMCA, 2015c). Parks Creek watershed management focuses too narrowly on floodplain management, and not on protecting headwater lakes (NBMCA, 2015c). The NBMCA and partners must be willing to support efforts to pursue more integrated management objectives (NBMCA, 1983a). Commitment and support for applying IWRM is discussed in the concluding section.

Integrated and coordinated planning on a watershed basis

The NMBCA administers programmes that range from sub-watershed to watershed scales. In pursing IWRM at various spatial scales, the NBMCA's responsibility is primarily achieved by fulfilling resource manager duties under the Conservation Authorities Act (1990). These activities normally involve hazard land mapping, alteration of shorelines, watercourses and wetlands, and delineating wetlands and critical habitats of species at risk. The NBMCA is also involved when functioning as a public commenting body under the Planning Act or Environmental Assessment Act, among other regulations. For example, the NBMCA has an agreement with member municipalities to review official plans to identify issues related to wetlands and critical habitats of species at risk. The municipal official plans guide future growth and development activities (NBMCA, 2015c).

A recent example of integrated and coordinated planning on a sub-watershed scale by the NBMCA includes the SPPs for five municipal drinking water systems in the North Bay-Mattawa Source Protection Authority (NBMCA, 2015c). As mentioned, the conservation authorities were delegated responsibility to coordinate the preparation and implementation of SPPs. The NBMCA appointed a Source Protection Committee that included municipal, provincial and community stakeholders. The area farmers elected an agricultural representative. First Nations were invited to participate, but were not involved in the SPP process. Narrowly scoped around municipal drinking water supplies, the SPPs identified threats to current and future drinking water supplies. The Source Protection Committee has various tools to eliminate or prevent risks to drinking water (e.g. restricted land use, land use planning, and education). In the case of the municipality of Callander, the increasing threat of cyanobacteria blooms in Callander Bay threatened its drinking water supply. The Source Protection Committee recommended that the Callander Issue Contributing Area receive special attention, so an education and outreach plan was initiated. The Issue Contributing Area Advisory Group chose a Restore Your Shore education and outreach campaign to reduce erosion and external phosphorus loading into Callander Bay. The project has been a success in the sense that the five municipalities, along with community stakeholders, are working together to address a drinking water quality problem in another municipality. Implementation began in spring 2015, and targets were met by fall 2015. The target was 30 shoreline property plantings, and 31 property owners participated in the first year of the programme. Over 1.8 km of shoreline frontage was planted with 508 trees, 2696 shrubs and 2595 perennial plants. The overall goal is 60 property plantings by end of Year 2 of the programme (NBMCA, 2016a). The watershed-scale integrated watershed management efforts are discussed in the next section.

Applying IWRM: lessons learned

As mentioned, the NBMCA is in the final stages of preparing its Integrated Watershed Management Strategy, to plan and coordinate the management of the next 20 years. This process included a reflection on the lessons learned over 44 years of implementing IWRM. This section describes the evolving practice of IWRM by comparing and contrasting the NBMCA's two watershed-scale strategic planning exercises. Many aspects of the two processes were similar. However, the process has been improved in many ways.

The first comprehensive watershed planning initiative began in the early 1980s. In 1982 the *Watershed Plan Volume I: Background Inventory Document* (NBMCA, 1982b) was published. The responsibility of the NBMCA was defined broadly to include watershed management, information and education, recreation, forestry, fish and wildlife and historical and archae-ological resources. Therefore, the document contained a comprehensive inventory of phys-ical and biological features, as well as some social and economic considerations. It provided a general assessment of the characteristics of the watershed. It also laid out the basis for identifying problems and conflicts, as well as for preparing sub-watershed studies. The pre-liminary Watershed Plan (NBMCA, 1983a) focused on problems and issues in 6 of the 16 sub-watersheds at that time. This approach is consistent with how the NBMCA implemented and defined comprehensive watershed planning:

> Comprehensive watershed management is not required in all basins. Such management is required in basins which have many or complicated land, water and resource management problems and issues. Comprehensive watershed management should be based on a plan, which has been developed to express policies, to co-ordinate physical, social and economic devel-opment and to initiate any regulatory or administrative measures necessary to implement or maintain the plan. (NBMCA, 1983a, p. 30)

The most recent watershed planning exercise follows an integrated watershed manage-ment approach. The NBMCA shares Conservation Ontario's (2012, p. 4) definition of inte-grated watershed management: "the process of managing human activities and natural resources on a watershed basis, taking into account social, economic and environmental issues, as well as community interests in order to manage water resources sustainably". Integrated watershed management "helps the NBMCA to identify and prioritize its manage-ment opportunities but it has implications for all stakeholders. Stakeholders share a common responsibility to cooperatively manage water and related resources features in fulfilment of riparian responsibilities shared by everyone" (NBMCA, 2013, p. 1).

Despite the use of different terms, the watershed planning exercises were similar in prepa-ration. An extensive inventory of watershed features is gathered, which then forms the basis of subsequent sub-watershed planning initiatives. Moving forward, the NBMCA will narrowly scope actions, monitoring and reporting relative to the threat. The priorities of the NBMCA are defined in consultation with member municipalities, provincial representatives and the public. However, the main distinguishing feature between the two watershed planning initiatives was the level of collaboration and community involvement. There is no reference to public involvement in the definition of comprehensive watershed management quoted above. The first watershed management plan was largely a desktop exercise, with little consultation with the public (NBMCA, 2015c). In subsequent sub-watershed planning activities, and the current integrated waters resource management strategic planning process, public consultation was a common practice. At the onset of any study, public involvement and project buy-in to the importance of the project are essential. That lesson was learned early, and ongoing provincial, municipal and public engagement is an important element of gaining support for local initiatives.

To facilitate further connections with community, municipal, provincial and federal partners, the NBMCA established a public liaison committee, for the first time, for its Parks Creek flood damage reduction environmental assessment study. This initiative involved more than public meetings. The committee provided more public accessibility and involvement in the process, relative to previous initiatives. The NBMCA viewed the committee as a success because it led to increased public input. This became the standard practice of the NBMCA in subsequent sub-watershed planning initiatives. Buy-in to the study process, recommendations, and implementation by all the stakeholders – interest groups, provincial and municipal agencies and the public – are necessary (W. Beckett, personal communication, 22 April 1992). This procedure has contributed to the successful implementation of many local initiatives, with measurable improvements in water quality and reduced flooding risk. The cumulative impact of these activities contributes to improving the overall state of some sub-watersheds.

The first watershed planning initiative was partly initiated at the request of the provincial government to create a five-year plan and budget. There were references to the lead agency and possible partnerships for specific issues, but there was no plan for monitoring or reporting progress, another key lesson that monitoring and reporting are necessary. The 20-year Integrated Watershed Management Strategy (IWMS) contains four 5-year plans (NBMCA, 2015c). The priorities include: collecting pertinent background information on pending IWMS actions; updating hazard land mapping; reviewing the broad-scale monitoring plan; and creating a wetlands and shorelines alteration policy. Each initiative identifies the partnerships required to meet the goals. After three years, the plan will be revisited to assess progress regarding implementation. Each initiative will be colour-coded according to process: green for on track, amber for behind schedule, and red for not advancing as expected (NBMCA, 2015c). The IWMS five-year plans may be modified, with regard to implementation progress or other unforeseen local or provincial needs.

The Technical Background Report (NBMCA, 2013) accompanying both watershed and sub-watershed planning initiatives focuses on physical, biological and chemical data, as well as economic and social data. The economic and social data include demographics, settlement, land use, labour force, industries and other elements. There is more physical, biological and chemical information available on the NBMCA administrative area compared to the 1980s Background Inventory (NBMCA, 1982b). The Assessment Report (NBMCA, 2015b) prepared for the SPP with Clean Water Act funding provided much of the scientific information in the 2013 Technical Background Report, including future climate, and surface water and groundwater interaction. Partnerships with post-secondary institutions help address water quality and quantity monitoring gaps. Formal partnerships need to be established in order to maintain a useful monitoring programme that encompasses environmental, social and economic indicators.

The IWMS identifies the specific environmental (e.g. protect provincial significant wetlands, protect cold water habitats), economic (e.g. diluted bitumen pipeline, sub-division development) and social (e.g. maintain public access to water) needs and issues for all 20 sub-watersheds in the NBMCA administrative area. As in the first watershed planning strategy, sub-watershed issues were prioritized based on water quality and quantity threats and recent trends. However, in 8 of the 20 sub-watershed assessments, the threats were assumed, because of insufficient information. The NBMCA IWMS programme priorities were linked to the traditional core functions and mostly involve gathering more data on meteorology and climate change, updated flood mapping, and expanding water quantity and quality monitoring. There is also a need to identify and map critical habitat of species at risk. The NBMCA has limited resources to undertake these initiatives, so partnerships with stakeholders will be essential to address these data gaps.

Is it worth the effort and expense to maintain an organization tasked with the responsibility of coordinating IWRM on a watershed basis? We think so. The case-study evidence above supports this assertion. Throughout 44 years, the NBMCA has proven to be an organization that can adapt to changing needs and circumstances at the provincial and local levels. Often provincial and municipal needs intersect at the sub-watershed scale, which leads to IWRM being applied more at the sub-watershed scale. As priority areas are addressed, the efforts shift to other immediate threats. The concluding section reflects on the NBMCA's structural and social barriers to implementing IWRM and solutions.

Conclusion

The NBMCA experience provides evidence that IWRM principles are being implemented. However, the transition towards IWRM is not a linear path; it is an evolving and turbulent process influenced by resources, institutional arrangements, commitment and social capital. The financial, human, technical and information resources are ongoing concerns that inhibit the successful implementation of any programme. The NBMCA has been able to successfully implement its core mandate by providing a valuable service to the municipal and provincial partners. However, the ability to undertake long-term planning is affected when project funding is limited in scope and time. Fewer summer students are hired, due to cuts to provincial and federal student employment programmes. The number of seasonal positions has dropped from nearly 50 in the early 1990s to 14 in 2016. The annual budget (CAD 90,000) for seasonal staff has not increased since the early 1990s. With less funding available from provincial and federal sources, the NBMCA relies on operating budget funds to hire season staff. Many post-secondary graduates gain valuable practical experience and apply their skills through summer jobs. Summer students led many sub-watershed flood and erosion studies in the late 1970s and 1980s (Shrubsole, Goodman, & Sullivan, 1980). The narrowing of core programmes and funding cuts in the 1990s reduced the NBMCA's ability to conduct regular environmental monitoring across the watershed. The NBMCA's ongoing collaborations with local post-secondary institutions and citizen scientists have increased its ability to monitor conditions. The sharing of resources – equipment, funds and people – is mutually beneficial. Also, the NBMCA has partnered with faculty at Nipissing University and Western University to share the costs of hiring students. A Natural Sciences and Engineering Research Council Collaborative Research and Training Experience (CREATE) grant has funded three research positions in the NBMCA. Also, the NBMCA regularly submits proposals to the Great Lakes Guardian Community Fund (Ontario Government, 2016a) and Ontario Trillium Foundation (Ontario Government, 2016b) to fund students and project activities.

The strategic initiatives and structural adjustments outlined above have presented opportunities and challenges. While the hazard land mapping and protecting drinking water supplies are narrowly scoped, these programmes indirectly contribute to building stronger relationships with municipal and provincial partners and the public. The NBMCA has been resilient with respect to the strategic initiatives and institutional changes largely because of the commitment of partner municipalities. The provincial funding cuts in the 1990s would have been much worse if the municipalities had not agreed to raise their municipal levy. This demonstrates commitment by the member municipalities to supporting the efforts of the NBMCA. The NBMCA is also building social capital by fostering community stewardship. The Issue Contributing Area education and outreach programme is using community-based social marketing principles to implement the Restore Your Shore programme (NBMCA, 2016a).

Applying IWRM requires more than one agency acting alone. The coordinating function of the conservation authority programme is an essential part of finding solutions that balance local and provincial interests. The efforts to foster community stewardship and a shared responsibility for natural resource management will further enhance opportunities to make choices that consider and balance environmental, economic and social values. However, these efforts alone will not overcome the challenges that the NBMCA faces. Conservation Ontario (2010) released a discussion paper, "Integrated Watershed Management: Navigating Ontario's Future", that outlines numerous barriers facing conservation authorities in Ontario that are relevant to the situation in the NBMCA. In the IWMS, the NBMCA (2015c) identified some possible remedies for these ongoing challenges. The expanding scope and complexity of watershed management is a challenge. At the tributary scale it is possible to scope problems narrowly to resolve isolated issues. For example, the harmful algal blooms occurring on occasion in Callander Bay can be reduced by controlling external and internal sources of phosphorus in the embayment.

The financial, human and information capacity of the NBMCA was compromised by the provincial cuts of the 1990s. While the NBMCA can cover most fields of expertise through existing staff, some areas, such as archaeology and hydrogeology, are externally sourced. In some instances, government experts or people from educational institutions fulfil the roles as required. The NBMCA continues to benefit from these informal collaborative partnerships. However, it is in the best interest of the NMBCA to formalize these partnerships through letters of support for research grants and other funding sources. This can also help address the data gaps. Endorsing research activity that contributes to the mandate of the NBMCA is mutually beneficial. Moving forward, data collection should be more strategically focused. There are opportunities to involve undergraduate and graduate students in monitoring and reporting activities to fulfil course requirements. Nipissing University offers students course credit for summer research assistantships. Conservation Ontario (2010) reported that in some conservation authorities the public failed to recognize the benefits of these programmes.

This did not appear to be an issue among the NBMCA staff, in part because of the ongoing collaborations and social marketing activities associated with the Source Water Protection Planning process.

While not noted in the Conservation Ontario (2010) report, engagement with First Nations is another enduring challenge among conservation authorities. The NBMCA must build stronger relationships with First Nations by finding alternative ways to engage First Nations in the planning and implementation of programmes. Regular updates and communications should be sent to First Nations with treaty rights to the watershed. The traditional knowledge of First Nations has not been actively sought. This could be an effective first step in building stronger relationships with First Nations, and strengthening the performance of IWRM.

Funding

This work was supported by the Natural Sciences and Engineering Research Council of Canada [grant number 448172-2014].

References

Biswas, A. K. (2008). Integrated water resource management: Is it working? *International Journal of Water Resources Development, 24*, 5–22.

Bullock, R., & Watlet, A. (2006). Exploring conservation authority operations in Sudbury, Northern Ontario: Constraints and opportunities. *Environments Journal, 34*, 29–50.

Chippewa Creek EcoPath. (2016). Adopt-the-creek-Program. Retrieved from http://chippewaecopath. ca/pages/Welcome.aspx

Conservation Ontario. (2010). Integrated watershed management – navigating Ontario's future. Retrieved from http://www.conservation-ontario.on.ca/media/IWM_OverviewIWM_Final_Jun2.pdf

Conservation Ontario. (2012). Watershed management futures for Ontairo – conservation authority whitepaper. Retrieved from http://www.conservation-ontario.on.ca/media/Watershed_ Management_Futures_for_Ontario_FINAL_0ct3.pdf

Dillon Consulting. (1975). *North Bay-Mattawa flood plain and fill line mapping*. Prepared by M.M. Dillon Limited Consulting Engineers and Planners. October, 1975 Report # 7284-01.

Giordano, M., & Shah, T. (2014). From IWRM back to integrated water resource management. *International Journal of Water Resources Development, 30*, 364–376.

Grigg, N. S. (2014). Integrated water resource management: Unified process or debate forum. *International Journal of Water Resources Development, 30*, 409–422.

Lawrence, P. L. (2011). Achieving teamwork: Linking watershed planning and coastal zone management in the great lakes. *Coastal Management, 39*, 57–71.

Millerd, F., Dufournaud, C., & Schaefer, K. (1994). Canada-Ontario flood damage reduction program-case studies. *Canadian Water Resources Journal, 19*, 17–26.

Mitchell, B. (1990). IWRM in practice: Lessons from Canadian experiences. *Journal of Contemporary Water Research and Education, 135*, 51–55.

Mitchell, B., & Shrubsole, D. (1992). *Ontario conservation authorities: Myth and reality* (p. 35). Waterloo: University of Waterloo, Department of Geography Publication Series No.

Mitchell, B., Priddle, C., Shrubsole, D., Veale, B., & Walters, D. (2014). Integrated water resource management: Lessons from conservation authorities in Ontario, Canada. *International Journal of Water Resources Development, 30*(3), 1–15.

North Bay-Mattawa Conservation Authority (NBMCA). (1982a). *North Bay-Mattawa conservation authority watershed plan inventory summary*. North Bay, Ontario: NBMCA.

NBMCA. (1982b). *Watershed plan volume 1: Background inventory document*. North Bay, Ontario: NBMCA.

NBMCA. (1983a). *The North Bay-Mattawa conservation authority watershed plan: Interim report*. North Bay, Ontario: NBMCA.

NBMCA. (1983b). *The North Bay-Mattawa conservation authority watershed plan: Interim watershed planning*. North Bay, Ontario: NBMCA.

NBMCA. (2013). *North Bay-Mattawa conservation authority integrated watershed management strategy: Technical background report*. Prepared by Stantec. Retrieved from http://www.nbmca.on.ca/site/docs/IWMS%20Technical%20Background%20Report.pdf

NBMCA. (2015a). *Integrated watershed management strategy*. Retrieved June 6, 2015, from http://www.nbmca.on.ca/site/indexa.asp?id=209

NBMCA. (2015b). *North Bay-Mattawa source protection area: Assessment report*. Retrieved from http://actforcleanwater.ca/uploads/Approved%20Assessment%20Report.pdf

NBMCA. (2015c). *North Bay-Mattawa conservation authority integrated watershed management strategy*. Prepared by Stantec. Retrieved from http://www.nbmca.on.ca/site/docs/NBMCA%20Integrated%20Watershed%20Management%20Strategy%20-%20DRAFT_Web.pdf

NBMCA. (2016a). Restore your shore program. Retrieved from http://actforcleanwater.ca/index.php?page=restore-your-shore-2

NBMCA. (2016b). Flood forecasting and monitoring program. Retrieved from http://www.nbmca.on.ca/news.asp?id=225

Nipissing University. (2016). Lake nipissing state of the basin report. Retrieved from http://lnsbr.nipissingu.ca/

O'Connor, D. R. (2002a). *Report on the walkerton inquiry: The events of May 2000 and related issues Part 1*. Toronto: Queen's Printer.

O'Connor, D. R. (2002b). *Report on the walkerton inquiry: A strategy for safe drinking water Part 2*. Toronto: Queen' Printer.

Ontario Government. (2016a). The great lakes guardian community fund. Retrieved from https://www.ontario.ca/page/great-lakes-guardian-community-fund

Ontario Government. (2016b). Ontario trillium foundation. Retrieved from http://www.otf.ca/

Pahl-Wostl, C. (2009). A conceptual framework for analysing adaptive capacity and multi-level learning processes in resource governance regimes. *Global Environmental Change, 19*, 354–365.

Pahl-Wostl, C., Jeffrey, P., Isendahl, N., & Brugnach, M. (2011). Maturing the new water management Paradigm: Progressing from aspiration to practice. *Water Resources Management, 25*, 837–856.

Patterson, James J. & Smith, Carl (2013). Understanding enabling capacities for managing the 'wicked problem' of nonpoint source water pollution in catchments: A conceptual framework. *Journal of Environmental Management, 128*, 441–452.

Rizvi, Z., Adamowski, J., & Patrick, R. J. (2013). First Nations capacity in Quebec to practice integrated water resources management. *International Journal of Water, 7*, 161–190.

Shrubsole, D. (1990). The evolution of public water management agencies in Ontario: 1946 to 1988. *Canadian Water Resources Journal, 15*, 49–66.

Shrubsole, D. (1996). Ontario conservation authorities: Principles, practice and challenges 50 years later. *Applied Geography, 16*, 319–335.

Shrubsole, D., Goodman, T., & Sullivan, C. (1980). *A preliminary investigation of the factors affecting erosion in the North Bay-Mattawa watershed*. North Bay, Ontario: NBMCA.

Ontario Statutes and Regulations

Building Code Act, S.O.. (1992). Retrieved from http://www.ontario.ca/laws/statute/92b23

Clean Water Act, S.O.. (2006). Retrieved from http://www.ontario.ca/laws/statute/06c22

Conservation Authority Act, R.S.O.. (1990). Retrieved from http://www.ontario.ca/laws/statute/90c27

Environmental Assessment Act, R.S.O.. (1990). Retrieved from http://www.ontario.ca/laws/statute/90e18

Planning Act, R.S.O.. (1990). Retrieved from http://www.ontario.ca/laws/statute/90p13

Integrated water resource management and British Columbia's Okanagan Basin Water Board

Natalya Melnychuk, Nelson Jatel and Anna L. Warwick Sears

ABSTRACT
This study examines successes and limitations of integrated water resource management (IWRM) for the Okanagan Basin Water Board (OBWB), a basin management entity in British Columbia, Canada. Effective governance, adequate financing and scientifically informed decision making are attributes contributing to the OBWB's IWRM success. OBWB's IWRM challenges include meaningful engagement of First Nations, public apathy towards water governance, succession planning for retiring professionals, and management authority limitations. Constraints on the OBWB's authority and perceived lack of need to formalize the IWRM approach will affect other local IWRM applications. The study adds a western Canadian example of basin management to IWRM practice.

Introduction

Integrated water resource management (IWRM) is generally characterized as holistic, respecting the many different uses and dynamics of water; interconnected, recognizing the relationships among water, other resources and land; sustainability-oriented, emphasizing economic, social and environmental goals; and inclusive, valuing the broad range of solutions that exist for dealing with water problems. Although there is significant academic debate on the practical application of IWRM (Blomquist & Schlager, 2005; Jeffrey & Gearey, 2006; Lautze, de Silva, Giordano, & Sanford, 2011), there are numerous examples of water management entities using elements of IWRM theory in practice (Cervoni, Biro, & Beazley, 2008; Grigg, 2008; Margerum, 1999). Examples of Canadian water management entities that are explicitly using elements of IWRM include Ontario's Conservation Authorities (Conservation Ontario, 2012), Saskatchewan's Water Security Agency (formally the Saskatchewan Watershed Authority, 2006) and Manitoba's Conservation Districts (Manitoba Water Stewardship, 2014). At the basin level, organizations that represent multiple groups and perspectives on basin-level management are a common way to practise IWRM. These organizations collect, monitor, manage and communicate water data; enforce laws; coordinate land and water management policies; resolve water conflicts; develop and carry out plans for water management; design, construct and manage water infrastructure; and educate the public about basin management (Global Water Partnership & International Network of Basin Organizations, 2009).

Given the different physical and human geographies across the globe and within each country, IWRM does not prescribe a single framework for achievement, and the application of IWRM functions has not been systematic across basins (Global Water Partnership Technical Advisory Committee, 2009). In general, key concerns facing the integration of water resource management in a basin include effective governance, adequate financing, and capacity (Hooper, 2003). These concerns relate to how decisions are made, the resources that are needed to sustain the management and decision-making process, and – in the context of this article – the ability of individuals and the organization as a whole to participate in informed decision making.

The Okanagan Water Basin Board (OBWB, www.obwb.ca) is a local managing organization in British Columbia (BC), Canada, that bridges three regional governments and 12 municipalities to coordinate water stewardship in the 8000 km^2 Okanagan Basin, with the advice and guidance of the range of affected actors. Although the OBWB and the province of British Columbia do not directly apply IWRM as a guiding framework, the ideals and approaches used by the OBWB illustrate the broad goals and objectives of IWRM, and the organization was probably influenced by IWRM theory when it was being established in 1970. By sharing the experience of the OBWB, this article will provide a practice-based example of basin management in western Canada based on the three key IWRM concerns.

This article discusses the context of BC and Okanagan water management, and reflects on the future challenges of the OBWB and limits of IWRM. These include the governance structure and processes of the organization, its financial model and its scientifically informed decision-making capacity.

The setting and management context

Canadian water management varies from province to province, often driven by ecological, political, economic and social circumstance (Bakker, 2006). In BC, water management responsibilities are divided among all orders of government – federal, provincial, regional, municipal and First Nations – and in some cases, among crown corporations, NGOs, the private sector, and one-off legislated arrangements like the OBWB. Political and basin boundaries are typically not aligned, and there is a trend towards the provincial government's further sharing its traditional responsibilities with (or devolving them to) municipal and regional governments, which in BC are called regional districts. Also, since the 1980s, BC has shifted to an emphasis on public participation and community-based approaches to environmental planning and management. However, unresolved First Nation treaties and debates on First Nations land and water rights have called into question the legitimacy of BC water management decisions (Day, Gunton, & Frame, 2003).

BC's main law on water management is the Water Sustainability Act (Province of British Columbia, 2014), which followed the Water Act (Province of British Columbia, 1996). In 2014, the hundred-year-old Water Act was modernized to update laws and policies to better reflect changing water demands due to growth of the BC population and economy. The resulting regulation changes are still under development and may alter many of the water management procedures in the province, including those in the Okanagan Basin. The Act follows the philosophy of 'one size does not fit all', favouring the application of different regulations in different areas. This continues the pattern of taking a patchwork approach to water management, leading to organizations with different mandates, organizational structures, levels of participation, and financing across the province (Brandes & Curran, 2009; Nowlan & Orr, 2010).

BC's varying approaches broadly fit into four categories (Nowlan & Bakker, 2007): (1) government-controlled decision making, with limited participation of non-government actors (e.g. Drinking Water Protection Plans); (2) NGO forums with extensive participation (e.g. Fraser Basin Council); (3) decision making among government actors, with limited participation of non-government actors (e.g. Langley Water Management Plan – Interagency Planning Team & Township of Langley, 2008); and (4) decision making with government and non-government actors across multiple interest groups (e.g. Okanagan Basin Water Board, Cowichan Watershed Board, Shuswap Watershed Council). As a result, basin organizations in BC range from voluntary to legally enabled and from citizen-based to top-down government-coordinated; and initiatives vary in duration (Morris & Brandes, 2013). Initiatives are significantly influenced by population, political, social and economic drivers; First Nations; and geographic and resource differences – as well as the political and financial attention different basins receive. As a result, this article's examination of IWRM and the OBWB is not reflective of the BC experience in general. However, the OBWB illustrates what is possible in BC, and the agency has substantially influenced the development of fledgling watershed governance structures elsewhere in the province.

Okanagan Basin

The Okanagan Basin is a narrow valley, approximately 200 kilometres long, in south-central BC (Figure 1). It is semi-arid, and water flows north-to-south through the Okanagan River, draining six main lakes and crossing the international boundary into the US as a tributary to the Columbia River. The area typically receives less than 30 cm of rain annually, in a precipitation gradient decreasing from north to south. The annual precipitation is highly variable, ranging from 100 million m^3 to over 1300 million m^3 annual inflow to Okanagan Lake (Summit Environmental Consultants Ltd, 2010). The Okanagan River is an important habitat for many plants and animals, including sockeye salmon and 23 species at risk (Canadian Okanagan Basin Technical Working Group, 2010).

The basin has been home to the Okanagan (Syilx) First Nations people since time immemorial. Their territory spans 69,000 km^2, spanning southern BC and into Washington State. The Syilx people now form eight member communities, which are represented by the Okanagan Nation Alliance (ONA, www.syilx.org). The Okanagan bands have not negotiated treaties with the province of BC; however, the ONA plays a significant role in water management in the basin, particularly concerning fisheries.

Major water issues in the basin include water scarcity, pollution, and invasive species control. Climate change exacerbates these issues through increased weather variability and extremes. Water scarcity in the basin results from both the natural dryness of the valley and a rapidly growing human population, with estimated average daily residential water demand of 675 litres per person (Okanagan Water Supply & Demand Project, 2011), compared to the Canadian average of 251 litres per person in 2011 (Environment Canada, 2014). Most human water use in the Okanagan is allocated to irrigation (55% for agriculture, 24% for residential irrigation, 7% for parks and golf courses), with a smaller fraction for indoor domestic and

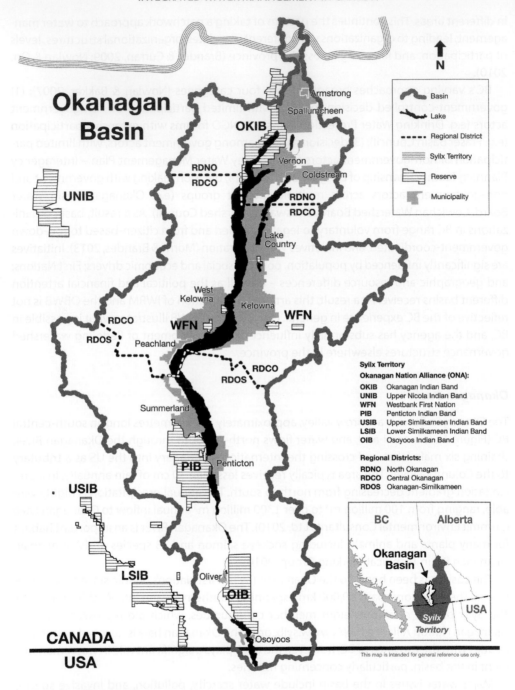

Figure 1. The Okanagan Basin.

industrial, commercial and institutional uses (14% of water allocations; Summit Environmental Consultants Ltd, 2010). Water scarcity concerns include increasing water demand, drought, and the timing of water availability throughout the year. In 2014, the estimated population of the three Okanagan regional districts together was 354,012, with an increasing annual

Table 1. The Okanagan Basin Water Board's Major Responsibilities (Okanagan Basin Water Board, 2009).

- Implementing basin-wide programs for water milfoil control, wastewater infrastructure funding, and water research and management to benefit all Basin residents
- Advocating and representing local needs to senior government planners and policy makers to protect Okanagan interests
- Providing science-based information on Okanagan water to local government decision makers and water managers for sustainable long-term planning
- Communicating and coordinating between government, non-government, universities and businesses to increase the effectiveness of water initiatives
- Building funding opportunities by providing leverage grants, securing external dollars and identifying cost-sharing partners to expand local capacity
- Partnering with government agencies and universities – especially in support of regional water science and monitoring

growth rate between 2013 and 2014 of 0.9–1.8% across the valley (BCStats, 2014). As a result of this population growth, water is likely to become more scarce in future years.

Point source and non-point source pollution are ongoing management issues, contributing to algal blooms and rooted aquatic plant growth. The main valley lakes are an important drinking water source, so municipal sewage and fertilizer runoff are a focus of management. Invasive species also receive attention, particularly Eurasian watermilfoil (*Myriophyllum spicatum*) control, and the prevention of other species such as zebra and quagga mussels (*Dreissena polymorpha* and *D. bugensis*).

The OBWB was formed in 1968 (legislated in 1970) to address valley-wide water concerns. In 1974, it was designated the local coordinating authority to address recommendations from the joint Federal/Provincial Okanagan Basin Study (Canada-British Columbia Consultative Board, 1974). Although the OBWB's original mandate included a valley-wide water leadership role, the initial overwhelming scale of managing the watermilfoil invasion and water pollution problems narrowed its focus for many years. From approximately 1973 to 2006, the agency focused almost exclusively on aquatic weed control and funding sewage infrastructure. Grass-roots concerns about rapid population growth, higher awareness of climate change, drought and forest fires led to a revitalization of the OBWB's mandate in 2006. The current mandate of the OBWB is to "provide leadership for sustainable water management to protect and enhance the quality of life and environment in the Okanagan Basin" (Okanagan Basin Water Board, 2010, p. 3).

Providing water leadership does not mean that the OBWB assumes all basin water management responsibilities. Other management entities involved in IWRM functions include federal and provincial agencies, First Nations fisheries, regional districts, municipalities, irrigation districts, research institutes and non-government stewardship and advocacy organizations. The basic responsibilities of the OBWB are summarized in Table 1. The direction of these responsibilities is informed by the advice of the Okanagan Water Stewardship Council (OWSC), a formally established technical advisory group to the OBWB. Functionally, the OBWB works as a bridge between the levels of government and other involved actors to find collective solutions for public water resource concerns.

OBWB experience

The organizational governance structure, financial model and science-based information philosophy of the OBWB have added to its reputation as a leading example of basin management in the province. Respectively, these factors have given the OBWB the jurisdiction, power of taxation, and multi-pronged science-based approach that have allowed the organization to be a voice of authority (without actual rule-making authority, as explained below) to help manage the Okanagan water resource.

OBWB governance

IWRM requires effective governance, which from a practical perspective stems from principles of good governance that form the basis of processes and institutions for decision making (Lautze et al., 2011). The six guiding governance principles of the OBWB are representation, service delivery, basin emphasis, collaboration, transparency and legacy (Okanagan Basin Water Board, 2010). These principles are the main tenets of the organization's management approach and illustrate the organization's governance strengths.

Representation

The OBWB is a collaborative governance institution comprised of elected public officials, ex officio members, and staff. The board of directors, primarily constituted of locally elected officials, is the main decision-making body that deliberates on options and seeks consensus to make decisions. The board is composed of three representatives from each of the three regional districts in the Okanagan Basin (Okanagan-Similkameen, Central Okanagan and North Okanagan), as well as representatives from the ONA, the Water Supply Association of BC and the OWSC. Together, these groups represent the approximately 350,000 people that live in the valley. The boundaries of the regional districts extend beyond the geographic limits of the basin, and the political jurisdictions within the basin include 12 municipalities, nested within the regional districts. The regional districts also govern unincorporated areas (Figure 1). The non-local government members vote and participate in all but the financial decisions of the board of directors. OBWB directors are appointed to represent the common water interests of all Okanagan citizens while providing the perspective of their independent constituency.

In addition to the board of directors, the OWSC ensures that a wide range of water interests can provide consolidated advice to the OBWB. For the 2013–2015 term, two dozen groups (Table 2) were represented on the OWSC. Each interest group appoints one person and an alternate. Each representative brings their own specific expertise and the position of their organization on water management to the table. The OBWB provides staff to the OWSC, and works to balance inclusive representation and a manageable organization structure.

Service delivery

The directors are responsible for meeting the OBWB's mandate: serving the basin as a whole. For its first four decades, the OBWB's services were mostly limited to watermilfoil control and the administration of a grant programme to fund sewage infrastructure improvements. When the OBWB's original, more integrated mandate was revived in 2006, it recognized that addressing more complex water issues requires a multi-faceted approach. Since then, the OBWB has expanded into education, awareness, advocacy, water modelling, data collection and management, and monitoring. Education initiatives include conferences and workshops

Table 2. List of involved interest groups in the Okanagan Water Stewardship Council (2013–2015).

- Agriculture and Agri-Food Canada
- Association of Professional Engineers and Geoscientists of British Columbia
- BC Agriculture Council
- BC Cattlemen's Association
- BC Fruit Growers Association
- BC Groundwater Association
- BC Wildlife Federation – Region 8, Okanagan
- Canadian Water Resources Association
- City of Kelowna
- City of Vernon
- Environment Canada
- Interior Health
- Ministry of Agriculture
- Ministry of Forests Lands and Natural Resource Operations
- Okanagan Collaborative Conservation Program
- Okanagan College
- Okanagan Mainline Real Estate Board
- Okanagan Nation Alliance
- Regional District of Central Okanagan
- Regional District of North Okanagan
- Regional District of Okanagan–Similkameen
- Okanagan Forest Sector
- UBC Okanagan
- Water Supply Association of BC

on water science, management and governance; social media; the Okanagan WaterWise outreach programme on valley-wide water conservation and protection; an active media presence; water by-law guidebooks for local governments; and a virtual library of publications and videos. Advocacy work has included policy recommendations on the BC Water Act modernization, land management of crown land in the upper basin, and prevention of invasive zebra and quagga mussels. Basin modelling, data collection, and monitoring provide a baseline of scientific data and ongoing monitoring to ensure credible, up-to-date information for decision making. Such modelling, data collection, and monitoring occur in a range of water management fields, such as hydrology (groundwater and surface water levels and extractions); climate change; watershed assessments; water management (governance, licensing, supply and demand); and agricultural, commercial, industrial and residential uses. Finally, the OBWB still maintains its basic water functions of watermilfoil control and funding for sewage infrastructure improvements, and maintains a small grants programme for water conservation and quality improvement by local governments and groups.

The diversity of the services provided by the OBWB constitutes a holistic approach, supporting water management by all organizations active in the valley, recognizing that there is no one solution for the sum of the valley's water problems, and that no one organization can do it all alone.

Basin emphasis

The OBWB takes an ecosystem-based approach that acknowledges and respects the different economic, social, spiritual and environmental values and uses for water across the region. The systematic nature of this principle has been fundamental in maintaining the support of each regional district and identifies closely with the Sylix (Okanagan First Nation) people's holistic and interconnected approach to water, land and natural resource management. As with all environmental management, there is an ongoing tension in how to balance competing demands and what trade-offs are reasonable. This has reinforced the value of the OWSC and related collaborations, where the OBWB seeks to include all key voices in discussions.

The call for basin-wide management came first in the 1960s from the Okanagan Pollution Control Council – the body that transitioned into the OBWB. From this call, the coordination and commitment of regional districts developed, under a philosophy of 'everybody pays, everybody benefits'. Sharing the financial responsibility of basin management means that programmes and projects are of value to the basin as a whole, and location-specific projects are distributed throughout the basin. The population of an area and perceived needs are considered when choosing location-specific projects.

The organization's basin-wide commitment means that the OBWB respects the upstream–downstream dynamic of the basin. For example, the OBWB places priority on cleaning wastewater through the funding of sewage treatment infrastructure. The goal is that any wastewater released will not harm drinking water or the aquatic environment downstream. Another example of basin-wide approaches is the OBWB's Okanagan WaterWise public education programme. The programme's motto, "one valley, one water", aims to educate and inspire basin residents about interconnected water use behaviours, and encourages change. Having one basin-wide communication programme creates consistency and economy of scale.

Collaboration

Central to collaboration in the Okanagan has been bringing water interest groups together at regular monthly meetings, for both the OBWB board of directors and the OWSC to discuss current water issues. At the OWSC meetings the intent is to provide independent advice and policy recommendations to the directors; at board meetings the intent is to make decisions based on these recommendations. The resulting output of these meetings includes collectively agreed-upon plans, policies and actions, such as the Okanagan Sustainable Water Strategy (2008), and specific advice on the OBWB's technical water science projects. Softer outcomes have also emerged – the monthly meetings build trust and cooperative learning among involved groups.

The benefits of collaboration are well recognized locally, as is reflected in the following comment by a senior provincial government bureaucrat:

> If I was to consider what the [OBWB] does right now, I would call it collaborative governance … it works extremely well … it acts as a catalyst to make all the different partners, whether or not they are federal governments, provincial governments, local governments and a whole other range of stakeholders working effectively together to solve whatever problem. (quoted in Jatel, 2013, p. 59)

Such sentiments are reiterated in the context of conflict resolution through collaborative engagement, as a senior federal bureaucrat states:

> There [are] linkages between all of the potential contenders … it really indicates the techniques [of water governance] in the Okanagan is doing a pretty good job of trying to defuse situations before they arise and [the OWSC] allows other people to see other perspectives … we have fish people talking on behalf of farmers and farmers talking on behalf of fish people. (quoted in Jatel, 2013, p. 59)

Collaboration among actors of diverse interests is long-term, energy-intensive work (Ansell & Gash, 2007). Both the OBWB's board of directors and the OWSC seek consensus, acknowledging the diversity of values in a collaborative process, and seek to ensure that everyone is heard before decisions are made. Collaborative learning is designed into the process so that all understand the proposals and issues, and speakers and guests are invited to share expert knowledge. This contributes to a base of common understanding. The participatory nature of both the board of directors and the OWSC helps integrate the processes and actions undertaken by the OBWB with the actions and policies of other levels of governments, as well as local water-sector organizations. Collaboration allows the OBWB to be a common voice for the valley and limits overlap with the actions and processes undertaken by other groups or agencies.

Transparency

Transparency concerns the visibility, clarity and accessibility of the decision-making processes, the rationale behind decisions, and other relevant information about the organization (Lockwood, Davidson, Curtis, Stratford, & Griffith, 2010). To the greatest extent possible, the actions and decisions of the directors, OWSC and staff are transparent and open. Systematic sharing and reporting back to the public, directly involved interest groups, and partners are formalized and practised as a governance strategy. Examples of actions that promote transparency include open meetings, monthly and annual reporting, and regular liaising with member local governments and community groups.

A comment from a senior provincial bureaucrat reflects the importance of transparency for the OWSC:

> The [OWSC] is fantastic ... there are minutes, they are open to the public ... people are part of the discussion ... how many other institutes [like the OWSC] are there around [British Columbia] ... the problem is not very many (quoted in Jatel, 2013, p. 58).

The clarity and visibility of the decision-making process underlie the OBWB's basin-wide support.

Legacy

The OBWB embraces the language of sustainable resource management – working to protect water resources for both current and future generations – and as a result focuses on medium- and long-term planning. The Okanagan Sustainable Water Strategy (2008) reflects long-term planning. Collaboratively developed by the OWSC, the strategy explains complex problems such as water pollution sources, data needs, and potential conflicts from population growth

Table 3. Guiding Principles for the Okanagan Sustainable Water Strategy (Okanagan Water Stewardship Council, 2008).

- Recognize the value of water
- Control pollution at its source
- Protect and enhance ecological stability and biodiversity
- Integrate land use planning and water resource management
- Allocate water within the Okanagan water budget in a clear, transparent, and equitable way
- Promote a Basin-wide culture of water conservation and efficiency
- Ensure water supplies are flexible and resilient
- Think and act like a region
- Collect and disseminate scientific information on Okanagan water
- Provide sufficient resources for local water management initiatives
- Encourage active public consultation, education, and participation in water management decisions
- Practice adaptive water and land management

and climate change. The strategy aligns with provincial and federal mandates; however, it is tailored to regional water issues. Table 3 outlines the main principles guiding the water strategy. At the broadest level, the strategy addresses surface and groundwater protection, water security, and good governance, informing planning in all areas. Complementing the water strategy, the OBWB has a five-year strategic plan that outlines shorter-term goals, objectives and actions to help implement the long-term strategy. The current strategic plan is for 2014–2019.

Basin-level water governance requires steady adaptive management, because the ability to make change in any given year is strongly affected by outside influences – including the weather (e.g. droughts and floods), the economy, election cycles, and policy by other levels of government. Formal reporting supports adaptive management as an active year-round process – proactively working towards the strategic plan goals, reactively adjusting to changes in external conditions, and opportunistically responding to proposals from partners that fit OBWB goals and objectives. Most importantly, reporting also makes the strategic planning process transparent. Annual reports are primarily used to communicate to local governments, stakeholders and partners, and every three years the OBWB undertakes a detailed review of water management projects and programmes.

OBWB finance

IWRM at the basin level needs adequate, reliable and sustained financing to support stewardship, infrastructure and the operation of the organization. Without consistent funding, it is difficult if not impossible to engage in long-term strategic planning and multi-year initiatives. By their very nature, watersheds need attention and commitment: there are no quick fixes when environment and human uses change through time. The basic governance structure of the OBWB is that of a shared purse: to collect and disburse funds over time to valley-wide priorities.

The OBWB was mandated under BC's Municipalities Enabling and Validating Act and given power of taxation through supplementary letters patent to the three Okanagan regional districts. This allows the OBWB to have consistent base funding for programme operations through annual property tax assessments on lands within the basin. Local base funding places the emphasis directly on local issues, and allows the OBWB to evaluate problems without being beholden to senior government priorities or being exposed to budget vacillations from an external source.

The OBWB is currently permitted to requisition up to $0.036 per $1000 assessed property value to cover the operational and project costs of its water management and watermilfoil control programmes. Any further increases in this formula must be agreed to by the electorate. The 2014/15 fiscal year budget for these programmes was $1.4 million (Okanagan Basin Water Board, 2014). This budget also includes overhead, equipment and staff, and the Water Conservation and Quality Improvement Grant Program to provide seed funds for small projects and leverage external grants. Grants are made to local governments, irrigation districts and non-profit organizations for small collaborative or basin-wide projects. The OBWB's sewage infrastructure grant programme has a separate, legislated ceiling of $0.21 per $1000, but the budget rarely exceeds $2 million per year. These larger infrastructure grants go to local governments to upgrade sewage treatment plants and help communities move from individual septic systems to community sewers, matching provincial infrastructure dollars.

The power to pool local dollars to protect, maintain and restore the basin makes the OBWB unique in the province. One of the agency's functions is to increase local access to resources in all possible ways. Along with grants, the OBWB provides in-kind services to local governments and non-profit organizations. Making expert staff available to all three regional districts also helps address shared problems that would have been too costly (if it were even possible) for one jurisdiction alone. Staff members also mentor the local governments on external grant proposals, write letters of support, and provide financial matches to secure and leverage additional external funding for specific projects. In 2012, the three regional districts requested that the OBWB grant $500,000 to the University of British Columbia Okanagan to support a research chair in water resources and ecosystem sustainability. The current chair's initiatives focus on research and partnerships at various scales concerning the governance, rights, uses and pricing of water locally and internationally. These funds triggered more than $1.5 million in contributions from the provincial government and the Real Estate Foundation of BC.

To add to its base funding, the OBWB regularly establishes funding agreements with senior governments for large water science and policy partnerships, which are described below. The OBWB has also received several grants from Natural Resources Canada to work on climate change adaptation, including workshops for local government staff.

The financial model of the OBWB is a key strength in its ability to consistently and effectively deliver its programmes. The base funds are not enough to cover the potential costs of all the work that could be done in the basin; however, they support core professional staff, and provide enough project and grant dollars to leverage the contributions of other funders. Having a cap on funding reduces fears and criticisms about the agency engaging in 'empire building', and also forces very careful strategic planning and emphasis on partnerships – which further reinforces the collaborative governance aspect of the OBWB (financial decisions must have broad support, because high-cost projects require buy-in from external partners). It would be possible, with changes to the legislation, to change the OBWB's financial cap and increase its financial capacity, but (in part because the cap is based on assessments that have increased over time) this has not been seen as necessary by the organization and its partners for many decades.

OBWB science information system

A central part of the legitimacy and credibility of the OBWB is its commitment to using the best available science and expertise of the community to inform decision making. For the OBWB, this has involved collecting data, supporting local water research, monitoring water resources and organizing a collaborative information system for sharing data and managing information. Multiple programmes and projects support this information system, including the GIS-based Okanagan Land Use Inventory and Water Demand Model (www.okanagan-water.ca), the Okanagan Water Supply and Demand Project (www.obwb.ca/wsd), the Okanagan Groundwater Monitoring Project (Jatel, Thomson, Graham, & Edwards, 2013), re-installing retired Water Survey of Canada hydrometric stations (Okanagan Hydrometric Network Working Group, 2008), creation of the BC Water Use Reporting Centre (www.bcwaterusereporting.ca), and numerous research studies.

In brief, the Water Supply and Demand Project concerns the assessment of Okanagan water availability (accounting for climate change and population growth) and includes modelling of groundwater, streamflows, environmental water needs and water use by balancing water supplies and demands using a computer accounting model. The Okanagan Groundwater Monitoring Project is a collaborative initiative among all levels of government to address monitoring gaps for priority aquifers in the Okanagan region. The re-installation of hydrometric stations in the basin is intended to increase ongoing data collection on streamflows where they are essential for understanding basin hydrology. The BC Water Use Reporting Centre is an online interface for large-volume water users in the Okanagan Basin to report their water use. Local research is often done in partnership with the University of British Columbia Okanagan and has recently included studies on endocrine disruptors in municipal wastewater, groundwater/surface water interactions, watershed studies, water governance, renewal of the Osoyoos Lake (see Figure 1 for location) operating orders under the Boundary Waters Treaty, and GIS mapping projects. Together, the information collected, organized and shared through these projects is helping decision makers make scientifically informed management decisions.

Table 4 categorizes which data collection and monitoring projects are undertaken by the OBWB and which are collected by other partner agencies and housed by the OBWB in a communal database and online library. Table 4 also highlights which projects are done by others and enabled through OBWB funding. The OBWB has compiled guides to help decision makers access and interpret these data, e.g. the *Local Government Guide to the Water Supply & Demand Project* (Okanagan Basin Water Board, 2011).

One of the most direct examples of how the OBWB's water science was used in decision making came with the renewal of the Osoyoos Lake Operating Orders, through the Boundary Waters Treaty, overseen by the International Joint Commission. During the process of renewing the orders, a technical recommendation was made to mandate certain flow levels in the Okanagan River at the outlet of Osoyoos Lake. The results of the Water Supply and Demand Project demonstrated that, with climate change, there was a high probability of recurring drought years, without sufficient water to meet Canadian environmental needs. As a result, the final orders did not mandate a given flow level, but reinforced the mutual intention of the US and Canada to protect fish habitat as a priority.

In another example, the province was considering the sale of recreation lease lots on upper-elevation reservoirs in the basin. Data and modelling of the current and future importance of these reservoirs, and the potential need to expand storage capacity in the future, swayed the government to abandon the sale plans. The OBWB is currently working with the ONA and the provincial government to use hydrology data and modelling to set in-stream environmental flow needs regimes for all major fish-bearing streams in the basin. These regimes will in turn influence future water licence allocations on these tributaries.

Although many instances exist where science-based information is effectively used in Okanagan decision making, there are still challenge areas. One area concerns climate change impacts on infrastructure, such as droughts and flooding. Decision makers in the Okanagan have been slow to use climate model results in formal assessments of infrastructure capacity. Despite the development of the Public Infrastructure Engineering Vulnerability Committee protocol (Engineers Canada, 2011), the standard practice in the Okanagan is still to add only 10% capacity as a buffer for climate extremes. The OBWB continues to work with local and senior governments, encouraging more detailed risk assessments using the latest science,

Table 4. Okanagan Basin Water Board Scientific Program and Project Data Management.

Project	Client/user	Website/access portal	OBWB Funded (partial or full)	OBWB Collected	OBWB Stored	OBWB Managed	Third party Collected	Third party Stored	Third party Managed
Data and information									
Water supply & demand	LG, WP, C, A, F, WL	www.obwb.ca/wsd	p	x	x	x			
Water demand model	LG, WP, C, A, F, WL	www.obwb.ca/wsd	p	x	x	x			
Okanagan Water Science Library	LG, C	www.bcwurc.ca	p	x	x	x			
BC Water Use Reporting Tool	LG, WP	www.bcwurc.ca	p	x	x	x			
Okanagan Water Database	LG	www.db.okanaganwater.ca	p	x	x	x			
Hydrometric stations (lake & river)	LG, P, C, A, F, WL	www.wateroffice.ec.gc.ca/	p				x	x	x
Hydrometric stations (non-network)	LG, C	(in progress)	p			x	x	x	x
Groundwater level & quality	LG, P, C, A, F, WL	www.env.gov.bc.ca/wsd/data_searches/obswell/	p				x	x	x
Upper reservoir levels & volumes	LG, WP, C, A, F, WL	www.bcwurc.ca	p	x	x	x			
Evaporation data	LG, C	www.bcwurc.ca	p				x	x	x
Policy and governance									
OBWB Governance Manual	LG, WP, P	www.obwb.ca/fileadmin/docs/obwb_governance_manual.pdf	f	x	x	x			
OBWB annual reports	LG, WP, P	www.obwb.ca/overview/annual-reports/	f	x	x	x			
OBWB Sustainable Water Strategy	LG, WP, P, C, A, F	http://www.obwb.ca/fileadmin/docs/osws_action_plan.pdf	f	x	x	x			
Outreach									
Okanagan WaterWise	WP	www.okwaterwise.ca	f	x	x	x			
Okanagan Basin Water Board corporate	LG	www.obwb.ca	f	x	x	x			
Water Stewardship Council	LT, WP, P	www.obwb.ca/about-the-council/	f	x	x	x			
Small Grant Program	LT, P	www.obwb.ca/overview-grants/	f	x	x	x			

User abbreviations:

LG Local government
WP Water purveyor
C   Consultants
A   Agriculture
F   Fisheries
WL Water licensing & management
P   Public

Notes: One strategic advantage of the OBWB's governance is the use of Regional District of the Central Okanagan data storage, security and management. Benefits include a high degree of security, ongoing management, data back-up and support protocols.

73

but uncertainty about insurance and liability laws is slowing uptake of climate data for infrastructure decisions.

The interface of science and decision making requires usable knowledge to develop the capacity of decision makers (Lemos, 2015). The science-based projects of the OBWB help facilitate this informed capacity development, and the OBWB recognizes the long-term nature of this task. Ongoing collaborative learning by the board of directors and the OWSC helps continually develop common knowledge among decision makers so that adaptive management can occur in the face of uncertain futures and environmental challenges.

OBWB strengths and limits, and further governance considerations

The OBWB, while not self-identified as an IWRM entity, follows many of IWRM's main tenets, as a basin management organization that considers economic, social and environmental needs. Geographic boundaries define its jurisdiction; the whole range of water actors are collaboratively engaged in the OBWB's planning and decision-making processes; and a multi-faceted scientific approach guides water management decisions that acknowledges, studies and considers land–water, surface–groundwater, and upstream–downstream interactions.

The OBWB provides a value-added service to local water management by the province, local governments and water purveyors. This takes a range of forms, from basin-wide water conservation education to basin-wide hydrology data and modelling. Most water utilities manage their own specific source and user area, and although their supplies may affect or be affected by other water users, without the OBWB's involvement there are often no formal ways to examine or respond to issues. For example, the OBWB developed the BC Water Use Reporting Centre to collect information on the difference between water licence allocations and actual water demand.

The OBWB supports the creation of more uniform local government water and development policy by creating draft model by-laws for water supply and water quality protection, and through grants for source protection assessments and sharing information on the structure, framework and process of these assessments among local government and water utilities. The OBWB is instrumental in leading, connecting, funding, facilitating and advocating valley-wide resilient water management. However, even with the accomplishments of the OBWB, challenges still exist, and improvements are possible.

The IWRM challenges in the Okanagan are both organizational and conceptual. Organizational challenges slow the OBWB's ability to achieve its vision, principles and objectives. One particular area that is in the process of change concerns First Nations engagement. Although the ONA holds seats on both the board of directors and the OWSC, questions exist about structural limitations on engagement in decision making, as First Nations government is a necessary partner in basin management (Okanagan Water Stewardship Council, 2008). Capacity limitations and risks to First Nations rights and title claims are often listed as reasons limiting First Nations engagement in collaborative management and governance processes outside of direct government-to-government initiatives (von der Porten & de Loë, 2013). As a result, structural governance changes may be needed to better include the ONA in OBWB processes. This may involve considering the direct relevancy and influence of the OBWB's programmes and projects, as well as appropriate engagement and communication strategies with the ONA. It may also involve considering opportunities for co-management between

First Nations people and local governments through further integration of OBWB processes, programmes and projects with First Nations water management – particularly as the ONA is now preparing its own water strategy.

Furthering IWRM in the Okanagan requires ongoing collaboration and adaptive management to mediate tensions among different human interests and environmental needs. Issues such as food security, residential development and the restoration of sockeye salmon runs in the Okanagan system all have water requirements and are long-term concerns. Collaboration helps prepare partners to deal with inevitable future trade-offs among interests so that conflict resolution is respectful of all parties (Emerson, Nabatchi, & Balogh, 2012). The OBWB must consistently deliver value and have steady communication with the public and involved partners to maintain and strengthen support for the collaborative approach.

Moreover, the Okanagan experience does not exist in a vacuum – awareness and anticipation of changes to the surrounding social context are necessary. For example, updates to BC's Water Sustainability Act, such as the introduction of groundwater licensing, will have implications for how the OBWB does business. The Act's new regulations have just come into effect in 2016; adapting to these regulatory changes in an integrated manner will require holistic assessment to understand how they influence human behaviour in other water management areas. For example, limiting groundwater licensing to priority areas could put increased pressure on other surrounding water sources.

Another external issue that requires adaptation by the OBWB is the retirement of water professionals, particularly those in government. This issue is well documented outside the water sector (MacKenzie & Dryburgh, 2003; Stam, 2009). The large baby-boom population contributed greatly to Canadian policy and infrastructure development during an era with an expanded public workforce. However, many of this generation are now retiring, and their knowledge, skills and expertise, acquired throughout their careers, may be lost. Capacity for water policy management and enforcement, as well as the expertise for maintaining aging water infrastructure, is at risk, as expert managers and technicians are leaving their positions without adequate succession planning (Grigg, 2006). The overall trend among senior government agencies is to downsize, with many positions not being replaced. Organizational preparation for the retirement of baby boomers is necessary so that decision making will continue to be informed by expert knowledge and an institutional history. Succession strategies may include improved documentation of information, as well as mentorship, knowledge sharing and training programs to better prepare the incoming workforce both within and outside government.

In addition to organizational improvements, the conceptual limits of OBWB's water management capacity when analyzing IWRM in the Okanagan are relevant. Despite the powers of taxation held by the OBWB, water management in the Okanagan is still primarily carried out by the province, the regional districts and local municipalities. For example, the OBWB exercises no direct authority, control or management responsibility over water levels, water licensing or groundwater. Numerous proposals have been made over the years to increase the OBWB's powers; however, governments have been reluctant to release control. Suggestions for change include adjusting the OBWB's structure, rebranding it as a Water Management Council and giving the resulting entity a range of powers to implement and coordinate basin-wide management policies; license water users; organize and manage upper-level reservoirs, control works and aquifers; and institute conservation practices such as water pricing (Nowlan & Bakker, 2007). However, it is unlikely the OBWB could assume such powers under existing political norms.

Greater formal authority is not necessarily a guarantee of greater influence or effectiveness. The OBWB is a convenor and facilitator, bridging local and senior governments as well as all parts of the water sector and stakeholders. These roles require an organization with the ability to bring people to the table and gain their trust. Pooled funding, shared priorities and information gathering and dissemination, as well as collaborative initiatives, give the OBWB 'soft power' to be productive in these roles in a contentious environment. If the agency were given regulatory responsibilities, there is a risk that the regulatory 'stick' would outweigh the benefits of the collaborative 'carrot'. Thus, the OBWB views the regulatory role as more appropriate for senior governments. Moreover, reservations also stem from limited resource capacity to take on such responsibilities, as well as normative concerns about excessive devolving of what the OBWB perceives as provincial responsibilities.

Incremental steps over the past decade have expanded the OBWB's mandate from water-milfoil control and sewage infrastructure funding. New responsibilities include actions such as managing collaboration among water-sector and agency partners through the OWSC, and assessment and research of surface and groundwater supply and demand and water quality studies. Although the OBWB's management responsibilities may continue to evolve, discussion of IWRM at an individual organization level must also include the extent of that organization's authority. In the case of the OBWB, where management is not limited to one organization alone, considering the vertical and horizontal integration of the organization with government and other organizations' policies is necessary.

Conclusion

IWRM is a means, not an end in itself (Global Water Partnership Technical Advisory Committee, 2009). As a framework, IWRM is not directly considered by the OBWB in its organizational development or its programmes and projects; however, significant overlap exists in the guiding principles and visions of IWRM and the OBWB. This overlap is not surprising, as IWRM bundles together a number of internationally recognized best practices, such as ecosystem-based management, water governance and adaptive management. Although this bundling may cause confusion regarding the application of IWRM, this all-encompassing approach allows context-appropriate practices to receive recognition as IWRM. This is the case for the OBWB.

The OBWB's achievements are part of an ongoing process of acting as a leader in basin-wide water management. The limitations and areas of improvement show the continuing work and considerations required. The OBWB experience underlines the importance of a basin-wide perspective that considers the integrated nature of all water sources and water users. For the Okanagan, basin water projects have often been best accomplished through inter-agency or inter-jurisdictional partnerships. The OBWB helps link the entities doing on-the-ground water management and creates synergies that would otherwise not be possible. The OBWB experience also highlights the importance of an agency that holds institutional knowledge that extends across jurisdictions and geographic areas. Continuity in the OBWB's staff helps reduce the impact of the retirement of water professionals and the turnover of local government. The OBWB brings the diverse group of government actors and stakeholders together using a stable funding source to collaboratively develop, implement and manage plans to satisfy all water interests using scientifically informed decision making.

References

Ansell, C., & Gash, A. (2007). Collaborative governance in theory and practice. *Journal of Public Administration Research and Theory, 18*, 543–571.

Bakker, K. (Ed.). (2006). *Eau Canada: The future of Canada's water*. Vancouver: University of British Columbia Press.

BCStats. (2014). *2014 sub-provincial population estimates*. Victoria, B.C.: Province of British Columbia.

Blomquist, W., & Schlager, E. (2005). Political pitfalls of integrated watershed management. *Society and Natural Resources, 18*, 101–117.

Brandes, O.M., & Curran, D. (2009). *Setting a new course in British Columbia - water governance reform options and opportunities*. Victoria, BC: The Polis Project.

Engineers Canada. (2011). *PIEVC engineering protocol for infrastructure vulnerability assessment and adaptation to a changing climate* (10th ed.). Ottawa, ON: Canadian Council of Professional Engineers.

Canada-British Columbia Consultative Board. (1974). *Canada-British Columbia Okanagan basin agreement*. Victoria, BC, Canada: Canada-British Columbia Consultative Board.

Canadian Okanagan Basin Technical Working Group. (2010). *Regional description*. Retrieved from: www.obtwg.ca.

Cervoni, L., Biro, A., & Beazley, K. (2008). Implementing integrated water resources management: The importance of cross-scale considerations and local conditions in Ontario and Nova Scotia. *Canadian Water Resources Journal, 33*, 333–350.

Conservation Ontario. (2012). *Watershed management futures for Ontario: Conservation Ontario whitepaper*. Newmarket, ON: Conservation Ontario.

Day, J., Gunton, T., & Frame, T. (2003). Toward environmental sustainability in British Columbia: The role of collaborative planning. *Environments, 31*, 21–38.

Emerson, K., Nabatchi, T., & Balogh, S. (2012). An integrative framework for collaborative governance. *Journal of Public Administration Research and Theory, 22*, 1–29.

Environment Canada. (2014). *Residential water use in Canada*. Retrieved from: https://www.ec.gc.ca/indicateurs-indicators/default.asp?lang=en&n=7E808512-1.

Global Water Partnership & International Network of Basin Organizations. (2009). *A handbook for integrated water resouces managment in basins*. Mölnlycke, Sweden: Global Water Partnership.

Global Water Partnership Technical Advisory Committee. (2009). *Lessons from integrated water resouces management in practice*. Stockholm, Sweden: Global Water Partnership.

Grigg, N. S. (2006). Workforce development knowland edge management in water utilities. *American Water Works Association Journal, 98*, 91–99.

Grigg, N. S. (2008). Integrated water resources management: Balancing views and improving practice. *Water International, 33*, 279–292.

Hooper, B. P. (2003). Integrated water resources managment and river basin governance. *Water Resources Update*, 12–20.

Interagency Planning Team & Township of Langley. (2008). *Water management plan*. (Draft report - 2nd version ed.) Victoria, B.C.: Ministry of Environment and Ministry of Agriculture and Lands.

Jatel, N. (2013). *Using social network analysis to make invisible human actor water governance networks visable: The case of the Okanagan Valley*. Master of Arts: University of British Columbia, Kelown, B.C.

Jatel, N., Thomson, S., Graham, G., & Edwards, D. (2013). *Okanagan groundwater monitoring project summary 2013*. Kelowna, B.C.: Okanagan Basin Water Board.

Jeffrey, P., & Gearey, M. (2006). Integrated water resources management: Lost on the road from ambition to realisation?. *Water Science and Technology, 53*, 1–8.

Lautze, J., de Silva, S., Giordano, M., & Sanford, L. (2011). Putting the cart before the horse: Water governance and IWRM. *Natural Resources Forum, 35*, 1–8.

Lemos, M. C. (2015). Usable climate knowledge for adaptive and co-managed water governance. *Current Opinion in Environmental Sustainability, 12*, 48–52.

Lockwood, M., Davidson, J., Curtis, A., Stratford, E., & Griffith, R. (2010). Governance principles for natural resource management. *Society and Natural Resources, 23*, 986–1001.

MacKenzie, A., & Dryburgh, H. (2003). The retirement wave. *Perspectives on Labour and Income, 4*, 5–11.

Manitoba Water Stewardship. (2014). *The conservation districts program: A manitoba success story*. Winnipeg, MB: Government of Manitoba.

Margerum, R. D. (1999). Integrated environmental management: The foundations for successful practice. *Environmental Management, 24*, 151–166.

Morris, T., & Brandes, O. M. (2013). *The state of the water movement in British Columbia: A waterscape scan & needs assessment of B.C. Watershed-Based Group*. Victoria, BC: Polis Project on Ecological Governance.

Nowlan, L., & Bakker, K. (2007). *Delegating water governance: Issues and challenges in the BC context*. British Columbia: UBC Program on Water Governance.

Nowlan, L., & Orr, C. (2010). *Brief on BC water act reform*. Vancouver: Watershed Watch Salmon Society.

Okanagan Basin Water Board. (2009). *2008–2009 annual report*. Kelowna, BC, Canada: Okanagan Basin Water Board.

Okanagan Basin Water Board. (2010). *Okanagan basin water board governance manual*. Kelowna, BC, Canada: Okanagan Basin Water Board.

Okanagan Basin Water Board. (2011). *Local government user guide: Okanagan water supply and demand project*. Kelowna, B.C.: Okanagan Basin Water Board.

Okanagan Hydrometric Network Working Group. (2008). *Hydrometric network requirements for the Okanagan Basin*. Retrieved from: http://www.env.gov.bc.ca/ecocat/

Okanagan Water Stewardship Council. (2008). *Okanagan sustainable water strategy: Action plan 1.0*. Kelowna, BC, Canada: Okanagan Basin Water Board.

Okanagan Water Supply & Demand Project. (2011). *Water use: Residents*. Kelowna, B.C.: Okanagan Basin Water Board.

von der Porten, S., & de Loë, R.C. (2013). Collaborative approaches to governance for water and Indigenous peoples: A case study from British Columbia, Canada. *Geoforum, 50*, 149–160.

Province of British Columbia. (1996). *Water act*. Victoria, BC: Province of British Columbia.

Province of British Columbia. (2014). *Water sustainability act*. Victoria, BC: Province of British Columbia.

Saskatchewan Watershed Authority. (2006). *2006–2007 saskatchewan provincial budget: Performance plan: Saskatchewan watershed authority*. Moose Jaw, Sask: Government of Saskatchewan.

Stam, C. D. (2009). *Knowledge and the ageing employee: A research agenda*. Haarlem, the Netherlands: European Conference on Intellectual Capital.

Summit Environmental Consultants Ltd. (2010). *Okanagan water supply and demand project: Phase 2 summary report*. Vernon, BC, Canada: Okanagan Basin Water Board.

The integrated watershed management planning experience in Manitoba: the local conservation district perspective

Colleen Cuvelier and Cliff Greenfield

ABSTRACT

Manitoba has abundant freshwater resources, and developing and implementing integrated watershed management plans is essential to ensure a healthy future. This article provides an assessment of progress in Manitoba since the early 1990s (Mitchell and Shrubsole, 1994) regarding integrated watershed management plans. It explains current conditions, including the structural framework, governance, public consultations and First Nations participation, along with examples of experiences, successes, failures, and lessons learnt. The Water Protection Act, proclaimed in 2006, empowered conservation districts to develop and implement integrated watershed management plans as the water planning authority, and represents the most significant change.

Introduction

Integrated watershed management (IWM) planning began in Manitoba in the 1990s (Mitchell & Shrubsole, 1994). This was a provincial government–led process engaging stakeholders and resulting in the creation of plans such as the Dauphin Lake Basin Management Plan in 1992, which was aimed at addressing water quality affecting the sports fishery around Dauphin Lake (Dauphin Lake Advisory Board, 1992). A new era of IWM planning in Manitoba began with the Water Protection Act in 2006, which gave non-government entities the authority to develop an integrated watershed management plan (IWMP) and defined the tools available to do so, along with a template to follow and guide the process. Lessons learnt from other jurisdictions in Canada and the pre-2006 processes in Manitoba have shaped the current IWM planning process in Manitoba. Today, the process is locally led by grass-roots, watershed-based conservation districts. The provincial government plays a key role by providing funding, technical assistance, oversight and logistical support, enabling grass-roots watershed planning. The evolution of the IWM planning process as experienced by conservation districts is the focus of this article. The viewpoint is from the experience of two conservation district managers, who have 35 years of combined experience working on programme delivery and development of several IWMPs. Consultations occurred with

30 colleagues from other conservation districts across Manitoba, as well as with provincial government staff working in this field.

Manitoba's Conservation Districts Program was born in the 1970s, modelled after the successful conservation authority programme in Ontario (Mitchell, Priddle, Shrubsole, Veale, & Walters, 2014). Municipalities partner with each other and the provincial government in the formation of a conservation district to deliver conservation programmes to local land-owners, develop watershed plans and manage natural resources. Between 1972 and 2008, 18 conservation districts were formed, covering the majority of agricultural Manitoba (Manitoba Conservation and Water Stewardship, 2015).

Manitoba's Conservation Districts Program

The four pillars of the Conservation Districts Program are: (1) grass-roots, community-based programme delivery, planning and management based on watersheds; (2) connection of upstream to downstream in watersheds; (3) financial support from a wide range of stake-holders; and (4) strong partnerships and cooperation. Local grass-roots leadership and local knowledge are combined with the technical expertise of stakeholder agencies, along with strong provincial support and oversight, to deliver on the vision of creating healthy water-sheds that support watershed residents, the environment and the economy for present and future generations.

Both the Conservation Districts Program and the IWM planning process have evolved and matured over the years to their present state, where the province's watersheds are improved by 18 conservation districts covering 85% of agricultural areas in Manitoba, and natural resources are managed with the help of locally directed IWMPs. Large urban areas have recently been integrated into some conservation districts, creating new opportunities and challenges, which are discussed later.

Although currently being updated, the Conservation District Act of 1976 governs the Conservation Districts Program, defining the fundamental structure and role of districts. Conservation districts are mandated to develop and follow schemes to manage natural resources in an area defined by boundaries established in the formation of a district. Manitoba is known as the Land of 100,000 Lakes (Manitoba Water Stewardship, 2003), and although it is blessed with abundant freshwater, natural resource management has been an ongoing challenge (Carlyle, 1983; Venema, Oborne, & Neudoerffer, 2010).

The Manitoba Conservation Districts Mandate Study concluded that the Conservation Districts Program creates two dollars of economic activity in the provincial economy for every dollar spent by the programme (FT-Ecologistics Limited, 1998). All conservation dis-tricts employ staff and maintain an office. Typically, a conservation district will employ full-time staff consisting of a manager, a financial administrator and a technician. In all, the 18 districts employ more than 50 full-time staff and approximately 20 seasonal post-secondary students. The local conservation district office typically provides watershed residents with information on available land and water management programmes and maps, delivers finan-cial incentive programmes, provides assistance in completing water rights licence applica-tions, and is an avenue to discuss water management challenges.

Within a conservation district, the organization is formalized into two groups: the main board and sub-district boards. Each member municipality appoints members to the sub-district board corresponding to the watersheds in their municipality. For each sub-district, one member is elected to serve as the chairperson of the sub-district committee. The main board consists of the chairpersons of the sub-district committees, along with one representative at large, appointed by the minister of conservation and water stewardship. The conservation district uses financial incentives, education and demonstration programmes to work with stakeholders to improve the health of the watershed through efforts prescribed in the IWMP.

The total budget for the Conservation Districts Program in 2014/2015 was CAD 10 million. The Manitoba government provided grants to the 18 conservation districts totalling CAD 5.2 million, plus an additional CAD 1.2 million in special funding. Participating municipalities contributed CAD 2.1 million, and external funding sources added CAD 1.6 million (A. North, personal communication, 12 May 2015).

From 1999/2000 to 2014/2015, municipal participation in the Conservation Districts Program increased by 79%. Over this same period, provincial grant funding increased by 100%, from CAD 2.58 million to CAD 5.2 million. This increase in provincial funding contributed significantly to supporting the growth and expansion of the Conservation Districts Program (A. North, personal communication, 12 May 2015).

Conservation Districts Program municipal levies

A conservation district levies funding from each of its municipal partners, based on the size of the provincial grant it receives. This cost-share is 3:1 (3 provincial, 1 municipal). The Conservation Districts Act (1976) provides the conservation district the right to request this levy from its municipal partners and defines how the levy is determined (based on the proportion of land assessment of the municipal land within the conservation district).

The Conservation Districts Program has brought on more municipalities and larger urban partners in the last several years, and this has been a challenge, since the programme was originally developed for rural agricultural areas. For example, the city of Brandon joined the Assiniboine Hills Conservation District in 2013. According to the 2011 census (http://www.economicdevelopmentbrandon.com/population), Brandon is the second-largest city in Manitoba, with a population of 46,061. Coming up with an appropriate cost-sharing arrangement has been difficult, although reconciling urban and rural differences is critical to watershed health and sustainability. In the case of the city of Brandon, raw water for the city's potable water system is withdrawn from the Assiniboine River, and this is blended with 25–40% groundwater during the spring freshet. Whether groundwater is used by itself or blended with other sources depends on the depth of the aquifer. The Assiniboine River, which runs through the middle of the city, poses a flood risk, so working with upstream stakeholders in a watershed context is sensible (N. Zalluski. personal communication, 4 May 2015).

Urban and rural municipal financial contributions to a conservation district vary, as do the natural resource concerns of the partners. In forming a partnership with an urban municipality, the natural resource needs are the highest consideration in developing the financial arrangement. As a result, no set formula is used; instead, a custom-designed cost-sharing arrangement is developed.

Benefits of integrated watershed management planning

Conservation districts are required to develop annual budgets outlining their works programme, major initiatives, operational expenses and suite of watershed programmes. They are also required to meet annually with the Conservation Districts Program Secretariat to discuss the draft budget and proposed activities. Linking IWMPs to actions with annual accountability keeps the plans off the shelf and active. The IWMPs identify specific actions, lead agencies and targets, with measurable indicators. An example is the Little Saskatchewan River IWMP, which identifies the involved conservation districts as leads to offer incentives for off-site watering and riparian fencing on watercourses and lakes (Little Saskatchewan River Conservation District, 2011). The approach to IWM planning by conservation districts is to move forward, get things roughly right, never be totally satisfied, and always strive to make progress and do better.

The Water Protection Act

With enactment of the Water Protection Act (2006), the provincial government defined the basic geographical framework of watershed units which sub-divided the 21 basins in the province. Figures 1 and 2 show the boundaries of the 18 conservation districts and the progress made on IWMP development. The watershed sizes for the IWMPs were based on what was thought to be manageable, with the sizes ranging from 1500 km^2 to 7000 km^2. Although previous resource planning based on watersheds had been done, the provincial government created a new scheme to complete IWMPs in all areas of Manitoba by 2015, using this new format. As of May 2016, 20 IWMPs had been completed, five were being prepared, and a few had yet to be started. All of the processes for IWMPs, whether completed or 'in the works', have been led by conservation districts. The number yet to be started will depend on the watershed boundaries to be used, and these will be negotiated with stakeholders in those watersheds.

In the legislation that defines IWMPs, the following highlights are important. A water planning authority is defined at the start of the planning process. As previously discussed, to date the water planning authority has been a conservation district board. The authority takes on the task of creating an IWMP, and one of its first priorities is to designate a project management team to oversee and complete the work. The project management team consists of local stakeholders – people who live and work in the watershed. It includes a conservation district manager, conservation district board members, watershed stakeholders, and one provincial staff delegated to write the plan and provide secretarial service. This provincial contribution speeds up the process. The project management team ranges in size from five to nine people. Although all watershed stakeholders are consulted and asked to participate, there is an extra degree of involvement for members of the project management team. Therefore, great efforts are made to include all interests in the planning process from the start. Not being inclusive can negatively affect participation by stakeholders during implementation.

The technical advisory group includes stakeholders called upon for their expertise in a particular aspect of the watershed, given their knowledge of the local landscape and resources. The project management team chooses the technical advisory group members. Not all aspects of every possible activity in a watershed are examined, as this is not viewed as practical. 'Integrated' does not mean 'all-encompassing'; certain aspects of a watershed are understood to have little or no consequence, and therefore investigating such aspects is not worth the effort. For example, if there are no mines or mineral extraction in a watershed, then this issue is not considered. Limiting the focus to pertinent issues speeds up the

Figure 1. Conservation districts are situated in most parts of agricultural Manitoba.

planning process. From start to finish, the IWM planning process currently takes on average two years to complete (A. North, personal communication, 4 March 2015).

As outlined in the Water Protection Act, a water planning authority must consider water quality standards, objectives and guidelines; water quality management zones and regulations; studies relating to water, land use, demographics and the environment; public input;

Figure 2. IWM progress in Manitoba watersheds.
Note: CD = conservation district. IWMP = integrated watershed management plan.

water management principles; provincial land use policies, development plans, and zoning by-laws; and any other relevant information. An IWMP must identify issues relating to the protection, conservation and restoration of aquatic ecosystems and drinking water sources; objectives, policies and recommendations respecting prevention, control and abatement of water pollution, surface water management and infrastructure maintenance; activities in water quality management zones and riparian areas; demand management, including planning for droughts and water shortages, and water supply; and emergency preparedness, to address accidents that may affect water. The plan must also make recommendations for adoption of water-related policies to be considered by local planning districts or municipalities, an implementation plan, ways to monitor and evaluate success, and a date for plan review.

Governance

Governance is defined in the enabling legislation that has followed and supports IWMP evolution. The governance approach balances local, grass-roots leadership, while also integrating stakeholder interests, along with provincial support. Conservation districts have developed IWMPs with assistance from government and non-government organizations; a plan is not operational until it receives provincial approval. Provincial oversight is necessary because some of the action items in the plan will become the responsibility of certain provincial government departments.

By sanctioning the finished plan, the provincial government is committed to its action items. An example is Action no. 1 of Goal no. 4 in the Pembina River IWMP: to 'promote a no-net loss of wetlands' (Pembina Valley Conservation District, 2011, p. 44). On the Prairies, wetland loss continues to be a concern for land managers, and there is an ever-increasing appreciation of the importance of this landscape feature. Consequently, much thought is being given to how to change this trend before most wetlands vanish (Manitoba Water Council, 2011). For the provincial government to agree to this goal, many provincial policies are implicated, and thus provincial departments need time to consider this goal and ensure that they can support it (A. North, personal communication, 18 May 2015). This provincial approval process increases the time needed to complete the plan, although many other aspects, such as limiting the scope to include only pertinent issues, have helped speed up the process. Past reviews of watershed management planning activities have concluded that the planning process needs to be sped up (Mitchell & Gardner, 1983).

A majority of municipal partners support the provincial government's having final approval. They do not view that action as overbearing and not trusting the local water authorities; instead, it is viewed as a seal of approval, committing the higher level of government to actions identified in the local, grass-roots-led plan (Barg et al., 2006).

Other jurisdictions across Canada follow a similar style of governance. For example, in British Columbia there has been a move to more local decision making, along with strong provincial oversight (Brandes & O'Riordan, 2014).

Structural framework for integrated watershed management plans

As the IWM planning process has evolved, the associated structures of the process have also changed. The Water Protection Act was enacted to provide the power and authority to create plans with legitimacy. Conservation district budgets must be tied to a plan's goals and targets, ensuring follow-through and a feedback loop connecting plan delivery agents to the funders. For things to change in the landscape, as directed by the IWMP, funding must be in place and secure, to create and maintain momentum. Momentum is important, as people and plans can become stale unless there is concerted and sustained effort to effect change through plan implementation.

Public consultation

Being integrated means consulting all parties with a connection to the watershed. Involving the public in watershed planning is a significant challenge. Low attendance at public forums is a concern. Diverse barriers exist, as diverse as the demographics of the general public, and thus need to be considered in design for public engagement (Huck, 2012). An example of a barrier to public engagement is that some people may be reluctant to get involved with something beyond their level of expertise because they expect the information discussed to be overly technical. Even though it is understood that water is a key to our survival and sustainability, the planning and protection of this resource may seem out of reach: part of the framework of government, or something requiring action, but by someone else. Other main barriers identified by Huck include general distrust of institutions, apathy and busy lives. These and other barriers need to be considered by watershed planners, who need to be innovative and provide many different forms and styles of interaction opportunities (Huck, 2012). An example of this may be to include information on watershed planning issues at a public festival or carnival and also to provide a method to receive feedback. In 2013, at a Canada Day (July 1st) celebration, anyone who wished to could fill out a watershed issue survey – and get three chances to throw a ball at a dunk tank and possibly soak a local celebrity.

Huck examined public engagement in two conservation district–led IWM planning processes. The Pembina Valley Conservation District led the Pembina River IWMP (2011), and the East Interlake Conservation District led the Netley-Grassmere IWMP (2011). Huck found that less than 1% of the watershed's population voluntarily participated in watershed meetings, implying that new and innovative ways are needed to engage the general public; it is important to try various methods of public engagement rather than just one or two. Local conservation district boards feel that articles and stories in local weekly or biweekly newspapers related to planning activities are the best method to reach the public, as opposed to open public forums or distribution of large and somewhat technical documents.

Huck noted that at all public information sessions for the two watershed plans, which were designed to provide and/or receive information and feedback, local municipal council members were always present. These councillors represent the public interest, and their presence does help legitimize the public engagement process. Barriers to public engagement are multifaceted and complex and something watershed planners need to improve on (Huck, 2012). Watershed planners need to understand the barriers and provide innovative, diverse and more numerous opportunities for public engagement (Huck, 2012).

First Nations participation

Many attempts have been made to involve First Nations in the watershed management planning process, but with minimal success, although indicators suggest improvement (Burt, 2014). Section 17(c) of the Water Protection Act (2006) states: 'In preparing a watershed management plan, the water planning authority must consult with … any band, as defined in the Indian Act (1985), that has reserve land within the watershed.' The water planning authorities to which this provision applies have contacted First Nations councils and invited participation in various ways.

Some successes with First Nations participation have been realized. The East Interlake Conservation District has succeeded in working with community members from the Peguis First Nation during development of the Fisher River IWMP in the northern Interlake area. The process started in 2013 and concluded with the creation of the Fisher River IWMP in 2015 (East Interlake Conservation District (EICD), 2015). The project management team chose the band's chief to act as chairperson. This cooperation was welcomed by all. It is difficult to pinpoint one action leading to this success, other than genuine willingness by all parties to collaborate. The Assiniboine Hills Conservation District partnered with the Swan Lake First Nation to fence Swan Lake's bison herd out of a locally important creek. This partnership was the result of relationship building through input and participation on the project management team in the development of the Central Assiniboine and Lower Souris River IWMP. Participation in both plans involved field tours, discussions of local traditions and hunting areas (Assiniboine Hills Conservation District, 2014), and many comments about the community's traditional way of life. The major benefits were sharing of knowledge, cooperation, and relationship building.

The Alonsa Conservation District has also been successful in engaging First Nation communities. This success is built on a relationship built on a spirit of sharing, respect, trust and communication. Over the past 10 years, the Alonsa Conservation District and the First Nation community have shared traditional knowledge and worked cooperatively to protect a special local area known as the Thunderbird Nest. When this area was brought to the attention of the conservation district by a local rancher, the conservation district immediately contacted the Ebb and Flow First Nation council to discuss a project to ensure the area's long-term protection. The groups worked together to place signage at the site and protect it for present and future generations. The conservation district performs annual maintenance of the trails around the site and facilitates school groups from as far away as Winnipeg to visit and appreciate the site.

The Swan Lake First Nation has been a strong advocate for a healthy watershed in the Pembina River system, because some of its property adjoins Swan Lake, which is part of the watershed, and the lake's water quality is often used as an indicator of ecosystem health. The Swan Lake First Nation is working with the Pembina Valley and Assiniboine Hills Conservation Districts and other watershed partners on improving the watershed, despite institutional barriers. For example, in the Conservation Districts Act, there is no reference to First Nations, and in the Indian Act (1985) the lands of a First Nation are identified as a federal responsibility. Other barriers are financial (limited funding for both conservation districts and First Nations), technical (lack of trained land and water managers in First Nations), and social/political (lack of awareness and support for watershed planning; Burt, 2014). The Conservation Districts Act is being revised, which promises to improve the situation, because the intent is to revise it in the spirit of Section 35 of the Canadian constitution, which stipulates 'the duty to consult' a First Nation in the watershed (Von de Porten & de Loe, 2010; Burt, 2014).

The hope is that in the near future, First Nations will be formally represented on conservation district boards and will have a significant role in the IWM planning process in Manitoba.

Source water protection assessments and plans

In Manitoba, source water protection is one part of IWMP development. A source water protection assessment is completed for each public water source, with recommendations for public, semi-public and private systems. The Office of Drinking Water (www.gov.mb.ca/conservation/waterstewardship/odw/) defines a public water system as a potable supply with 15 or more service connections. Semi-public water systems are public or private systems with fewer than 15 service connections or a public facility such as a school or hospital with its own water supply. Most source water protection plans in Manitoba include recommendations specific to the public water systems as well as private systems. The recommendations for private systems are generally directed towards wellhead management, given that most watershed plans are developed in rural areas. This approach to source water protection was legislated in the Water Protection Act. Before then, Manitoba had used the above approach for selected aquifers used for drinking water, namely in the Winkler Aquifer Management Plan (Manitoba Conservation and Water Stewardship, 1997), Oak Lake Aquifer Management Plan (Manitoba Conservation and Water Stewardship, 2000), and Assiniboine Delta Aquifer Management Plan (Manitoba Water Stewardship, 2005). Golder and Associates created the template for source water protection assessments (A. North, personal communication, 18 May 2015).

Protecting water at its source, before it arrives at treatment facilities, is a preventative approach. It is normally less expensive and more ecologically responsible to prevent contamination of source waters than to remediate water quality in treatment facilities. Source water protection strengthens the integrated approach to watershed management. When water is protected at its source, the streams, lakes and aquifers from which water is drawn are protected. This approach requires consideration of activities in the entire watershed, or at least in the area contributing to that stream, lake, or groundwater system.

Source water protection extends beyond protecting water in the watershed and can include protective measures for the drinking water treatment system and the distribution system carrying treated water to homes. Manitoba's planning process is focused on protecting water at its source or before it reaches municipal distribution systems.

Surface water management planning

In April 2008, the Manitoba Ombudsman (2008) released a 'Report on the Licensing and Enforcement Practices of Manitoba Water Stewardship', which recommended that the Manitoba government involve conservation districts in all drainage licensing decisions. The approach to surface water management planning in Manitoba subsequently developed as a high-level, watershed-scale assessment to enable coordinated water management for conservation districts and their watershed partners across the province.

Conservation districts are uniquely positioned to coordinate surface water management activities because:

- The districts are based on watershed boundaries. They operate with consideration of the entire watershed; uphold watershed principles, such as respecting the connectivity within a watershed and the rights of downstream landowners; and guide decision makers to ensure that decisions balance ecological, economic and community values.
- The districts are the preferred water planning authority in Manitoba and a platform for development and implementation of IWMPs. The Manitoba IWM planning process depends heavily on the support and engagement of water-related stakeholders, including those with surface water infrastructure responsibilities. This multi-stakeholder

approach is important for creating surface water management plans that are inclusive (including input from both provincial and municipal jurisdictions), watershed relevant, and respectful of interdependent needs such as water quality, aquatic ecosystem health and assimilative capacity of the watershed as a whole.

- The districts are managed and operated by local people. Understanding local issues and solutions has been key to creating plans focused on issues to maintain a healthy economy and supportive ecosystems.

- The districts use a surface water management plan to help make decisions about water management and protection. The process of developing the plan builds capacity in the conservation district board and staff and their government counterparts to improve understanding of the landscape characteristics unique to their watershed and the surface water management issues resulting from these natural and human-shaped landscape features.

Developing surface water management policies

Using watershed information gathered through technical submissions and stakeholder consultations (e.g. soils or agricultural capability, topography, and municipal land assessment data), management zones in areas with physiographic similarities are created (Manitoba Conservation and Water Stewardship, n.d.). A management zone is a unique geographic area within a watershed that the plan will focus on and it may be defined by land with similar soil types, land use, topography or a sensitive area such as the riparian area along a stream reach. A statement of intent is created for each management zone, along with policies and objectives reflecting that intent. The overall area and its future potential regarding downstream impacts (not just existing problems) are considered. Future surface water management decisions in that zone should reflect these policies and objectives.

Existing problems or identified projects in the zone are also outlined. Actions specific to one site within the zone are developed. Each specific action must fit the policies and objectives of the intent of the area for surface water management. The majority of these specific actions are determined through public consultations and individual meetings with municipal partners regarding their issues with surface water management in terms of maintaining infrastructure. In addition, most conservation districts have had various studies completed through their regular programming, including riparian and aquatic ecosystem health assessments. These assessments identify areas of threatened riparian health, generally identified as low, moderate or high, and may also identify sites where barriers to fish passage exist. Although the recommendations from these reports may help frame the policies and objectives of the area, they also outline specific sites within riparian and aquatic ecosystems to be repaired.

Integrated watershed management plans linking to development planning

In Manitoba, the Planning Act (2005) defines a development plan as a document which sets out the long-term goals, objectives and policies which will guide the future use and development of all land within a planning district or municipality. These plans have a 10-year life cycle, with a refresh after five years if necessary, as determined by the responsible provincial minister. The Planning Act and the Water Protection Act reference each other's planning process and recognize the need to link them. This integration is relatively informal, and not an absolute requirement. Beukens (2013) suggests that a more formalized collaborative process be considered. Although development planning considers much more than water management and land development, water planning authorities want to ensure that the connection of land use planning to watershed health is addressed. From 2008 to 2010, both the Pembina River IWMP and the Netley-Grassmere IWMP pushed to provide direct links to actions to be included in development plans in each watershed. This required consultation and correspondence between the conservation district's IWM planning effort and the planning district. IWMPs created after 2010 have taken further steps to integrate direct links to water- and watershed-relevant development planning. As the spatial boundaries of the two processes are different, this can present a challenge. The Pembina River IWMP is based on the watershed, and a planning district is municipally based. This resulted in eight separate planning documents needing to be reviewed to determine how the IWMP's recommendations fit into the development plan (Pembina Valley Conservation District, 2011, p. 34).

Implementation of integrated watershed management plans

Allocating and focusing limited resources to achieve measureable goals is what an IWMP is all about. For example, the Pembina River IWMP implementation period is from 2011 to 2021. Estimates indicate that CAD 5,000,000 will be spent by multiple stakeholders to achieve plan goals. In watersheds across Manitoba, new tools are being investigated and tried, including reverse tendering, LiDAR Internet tools, CanSWAT (computer modelling simulating the watershed landscape and beneficial management practices' impact on nutrient transfer), Alternative Land Use Services, and the Investment Framework for Environmental Resources (INFFER, http://www.inffer.com.au), a tool used to quantify environmental investments and link them to measurable outcomes. These tools and others are being considered to increase the cost effectiveness of plan implementation and achieve measurable target goals.

In the Pembina River IWMP implementation process, INFFER was used to quantify the cost-effectiveness of implementing the recommendations in the source water protection plan. The proposed actions were deemed to have a 20:1 benefit–cost index. A benefit–cost index is similar to a benefit–cost ratio except that it explicitly includes environmental values. This means that for every dollar of investment of a particular action there was a corresponding CAD 20 benefit. INFFER was initially developed in Australia. These tools, although not perfect, provide local boards with extra confidence regarding their fiscal decisions.

The latest federal/provincial/territorial agricultural agreement, Growing Forward 2 (Manitoba Agriculture & Rural Development, 2013), has provided funding for adoption of beneficial management practices in watersheds across Manitoba. Projects are ranked based on criteria such as how the activity will address an issue identified in an IWMP (Manitoba Agriculture & Rural Development, 2013). This process is bringing CAD 750,000 per year for a five-year period (2013–2018) into the Conservation Districts Program's IWMP implementation.

With implementation, there is also the need for monitoring and developing indicators that show whether actions are having an impact on watershed health. Presently, conservation districts focus on plan implementation, with little spent on monitoring. This is illustrated in the Manitoba Conservation Districts Program 2013–14 Annual Report (Manitoba

Conservation and Water Stewardship, 2014a). Conservation districts hope that new tools and indicators will be developed to track success. Computer modelling of beneficial management practice impacts may be able to do just that.

Conclusions

The IWM planning process continues to evolve. Each watershed is treated as a unique entity. Thus, a cookie-cutter or standard approach is not used. The examples of planning and implementation in Manitoba's diverse watersheds show flexibility, adaptation, creativity and imagination. In Manitoba, the Conservation Districts Program has led all IWM planning efforts since the passing of the Water Protection Act. Districts are effective in this role thanks to their grass-roots connections, their ability to adapt to change and their capacity to continue developing and maintaining collaboration and partnerships among municipal and provincial governments, local watershed residents and stakeholders. By utilizing conservation districts as water planning authorities, Manitoba has connected the local watershed knowledge and network to technical experts in the provincial and federal governments as well as partnering organizations, to make locally relevant, science-based IWMPs which are being systematically implemented.

Aspects needing improvement include the time it takes to create an IWMP. Plans have taken two or more years to complete, and effort being made to speed up the process. Speeding up the IWMP process is a balancing act between being thoroughly consultative, comprehensive and integrated enough, and being current and dealing with issues in a timely manner. Typically, the planning process does not hinder implementation, as the plan feeds into continual implementation by the conservation districts and others. To minimize the time required to complete the process, there has been some limiting of IWMP scope to deal with high-priority issues facing the watershed, as determined by stakeholders throughout the process. For example, if gravel pit mining and rehabilitation is not an issue in a particular watershed, then this issue is not investigated and reported on, which can save time.

There is still less public participation than desired. In one study that assessed two recent IWM planning processes, each recorded less than 1% of the population as engaged in the process (Huck, 2012). More positively, several IWM planning processes have engaged and are engaging First Nations. In the Fisher River IWMP, the band chief is chairing the project management team (EICD, 2015). Manitoba thus has a vision for a sustainable future, with IWM as a strong component to facilitate such a future (Manitoba Conservation and Water Stewardship, 2014b).

Disclosure statement

No potential conflict of interest was reported by the authors.

References

Assiniboine Hills Conservation District. (2014). *Central Assiniboine and Lower Souris River integrated watershed management plan*. Retrieved from https://www.gov.mb.ca/conservation/waterstewardship/iwmp/central_assiniboine/central_assiniboine.html

Barg, S., Bhandari, P., Drexhage, J., Kelkar, U., Mitra, S., Swanson, D., …Venema, H. (2006). International institute for sustainable development and the energy and resources institute and adaptive resource management ltd. Canada. 2006. *Designing policies in a world of uncertainty, change, and surprise. Chapter 8 Adaptive policy case study: analysis of Manitoba's conservation district policy.* Retrieved from http://www.iisd.org/pdf/2006/climate_designing_policies.pdf

Beukens, R. (2013). *Connecting watershed and land use planning in Manitoba: Exploring the potential of collaboration as a form of integration* (A practicum submitted to the Faculty of Graduate Studies of The University of Manitoba in partial fulfillment of the requirements of the degree of Master of City Planning, Department of City Planning). University of Manitoba, Winnipeg.

Brandes, O., & O'Riordan, J. (2014). *Decision-Makers' brief: A blueprint for watershed governance in British Columbia, Polis project on ecological governance.* University of Victoria, Polis Institute. Retrieved from http://poliswaterproject.org/sites/default/files/WaterDecision-3b1-hi.pdf

Burt, M. (2014). *First nations involvement in source water protection in Manitoba* (A thesis submitted to the Faculty of Graduate Studies, University of Manitoba in partial fulfillment of the requirement of the degree of Master of Natural Resources Management, Natural Resources Institute, Clayton H. Riddell Faculty of Environment, Earth, and Resources). University of Manitoba, Winnipeg.

Carlyle, W. J. (1983). Agricultural drainage in Manitoba: The search for administrative boundaries. In B. Mitchell & J. Gardner (Eds.), *River basin management: Canadian experiences* (Ch 21, pp. 278–295). Waterloo, ON: Published by the Department of Geography, Faculty of Environmental Studies, University of Waterloo.

Dauphin Lake Advisory Board. (1992). *Basin management plan.* Retrieved from http://www.gov.mb.ca/conservation/waterstewardship/questionaires/surface_water_management/#introduction

East Interlake Conservation District (EICD). (2015). Fisher river integrated watershed management plan. Retrieved from http://www.eicd.ca/main.asp?cat_ID=23

FT-Ecologistics Limited. (1998). *Manitoba conservation districts mandate study.* Brandon, Manitoba: Manitoba Water Services Board, October.

Huck, D. (2012). *A question of perspective: Opportunities for effective public engagement in watershed management planning in Manitoba* (A Thesis submitted to the Faculty of Graduate Studies of the University of Manitoba in partial fulfillment of the requirement of the degree of Master of Natural Resources Management). University of Manitoba, Winnipeg.

Little Saskatchewan River Conservation District. (2011). *Little Saskatchewan River integrated watershed management plan.* Retrieved from http://littlesaskatchewanrivercd.ca/wp-content/uploads/2011/01/LSR-IWMP.pdf

Manitoba Conservation and Water Stewardship. (1997). *Winkler Aquifer Management Plan.* Retrieved from http://www.gov.mb.ca/waterstewardship/reports/acquifer/winkler_aquifer_mp.pdf

Manitoba Conservation and Water Stewardship. (2000). *Oak Lake aquifer management plan.* Retrieved from http://www.gov.mb.ca/waterstewardship/reports/acquifer/oak_lake.pdf

Manitoba Conservation and Water Stewardship. (2014a). Manitoba conservation districts program 2013-14 annual report. Retrieved from http://www.gov.mb.ca/conservation/waterstewardship/misc/cd_annual_rpt_2013_14.pdf

Manitoba Conservation and Water Stewardship. (2014b). *Tomorrow now, Manitoba's green plan 2nd edition.* Retrieved from http://www.gov.mb.ca/conservation/tomorrownowgreenplan/pdf/tomorrownow_v2.pdf

Manitoba Conservation and Water Stewardship. (2015). *Manitoba's conservations districts.* Retrieved May 23, 2015, from http://www.gov.mb.ca/waterstewardship/agencies/cd/index.html

Manitoba Conservation and Water Stewardship. (n.d.). *Surface water management strategy.* Retrieved from http://www.gov.mb.ca/conservation/waterstewardship/questionaires/surface_water_management/#introduction

Manitoba Water Council. (2011). *Seeking Manitobans' perspectives on wetlands, what we heard.* Retrieved from http://www.manitobawatersoucil.ca

Manitoba Water Stewardship. (2003). *The Manitoba water strategy.* Retrieved from http://www.gov.mb.ca/waterstewardship/waterstrategy/pdf/index.html

Manitoba Water Stewardship. (2005). *Assiniboine delta aquifer management plan.* Retrieved from http://gov.mb.ca/conservation/waterstewardship/reports/acquifer/assiniboine_delta_aquifer-mgmt_plan.pdf

Mitchell, B., & Gardner, J. (eds.). (1983). *River basin management: Canadian experiences. university of waterloo*. Waterloo, Ontario: Department of Geography Publication Series No. 20.

Mitchell, B., & Shrubsole, D. (1994, October). Canadian water management: Visions for sustainability, Canadian water resources association, chapter 3 Experiences and Innovations in Canada by Jurisdiction (pp. 21–26).

Mitchell, B., Priddle, C., Shrubsole, D., Veale, B., & Walters, D. (2014). Integrated water resource management: Lessons from conservation authorities in Ontario, Canada. *International Journal of Water Resources Development, 30*, 460–474. doi:10.1080/07900627.2013.876328.

Ombudsman, M. (2008). Report on the licensing and enforcement practices of Manitoba water stewardship. Retrieved from http://www.ombudsman.mb.ca/uploads/document/files/report-licensing-enforcement-water-stewardship-2008-en.pdf

Pembina Valley Conservation District. (2011). *Pembina river integrated watershed management plan. / Action #1 of Goal #4 /*. Retrieved from http://www.pvcd.ca/PembinaRiverIWMP.pdf

Roseau River Watershed Plan. (2007). Retrieved from http://srrcd.ca/wp-content/uploads/2013/12/Roseau-River-Watershed-Plan.pdf

Venema, H. D., Oborne, B., & Neudoerffer, C. (2010). *The Manitoba challenge: Linking water and land management for climate adaptation*. Winnipeg, MB: International Institute for Sustainable Development.

Von de Porten, S., & de Loe, R. C. (2010). *Water challenges and solutions in first nations communities*. Waterloo, ON: University of Waterloo, Water Policy and Governance Group.

Statutes, regulations and programmes

Manitoba Agriculture and Rural Development. (2013). Retrieved from http://www.gov.mb.ca/agriculture/growing-forward-2/

The Indian Act. (1985). Revised statutes of Canada, c. I–5. Retrieved from https://www.canlii.org/en/ca/laws/stat/rsc-1985-c.../rsc-1985-c-i-5.html

The Water Protection Act. 2006. C.C.S.M. & W65. Retrieved from http://web2.gov.mb.ca/laws/statutes/ccsm/w065e.php

The Conservation Districts Act. (1976). Continuing consolidation of statutes of Manitoba, Chapter C175. Retrieved from http://web2.gov.mb.ca/laws/statutes/ccsm/c175e.php

The Planning Act. (2005). Continuing consolidation of the statutes of Manitoba, Chapter P80. Retrieved from http://web2.gov.mb.ca/laws/statutes/ccsm/p080e.php

Applying integrated watershed management in Nova Scotia: a community-based perspective from the Clean Annapolis River Project

Levi Cliche and Lindsey Freeman

ABSTRACT
This article examines integrated watershed management in the Annapolis River basin in Nova Scotia from the perspective of a community-based watershed organization. It draws on the experiences of the Clean Annapolis River Project (CARP) to provide a case study of the financial, institutional, human, political and social capacity of a small non-governmental organization in implementing integrated watershed management. CARP's guiding principles of utilizing science, leadership and community engagement to achieve ecologically healthy watersheds align with an integrated watershed management approach. Using examples of CARP's programming and projects, this article describes the successes and challenges encountered in the implementation of community-based integrated watershed management.

Introduction

Integrated water management has become an increasingly important approach to managing water resources and the impacts of human activities on watersheds. Substantial cuts to funding for environmental programmes and the increasing complexity of issues facing ecosystems have reduced governmental ability to adequately monitor and manage local watersheds (Conrad & Daoust, 2008). Community-based groups can offer a necessary link to help address gaps in watershed management by providing a cost-effective, local perspective for management needs that are not constrained by political boundaries. The following article provides an in-depth case study of the experiences of a small Nova Scotian community-based watershed group in adopting an integrated approach to watershed planning and management. It examines how the guiding principles of the Clean Annapolis River Project (CARP) align with an integrated watershed management approach, and discusses the successes and challenges in monitoring and managing a watershed from the perspective of a local community-based watershed organization.

Integrated water resource management in Nova Scotia

Management of water resources and watersheds in Nova Scotia to date has been fragmented, and has lacked an adequate guiding policy framework (Sharpe & Conrad, 2006). Other provinces, such as Ontario and Manitoba, have increased integration of water resource management through the creation of conservation authorities and conservation districts, respectively, which utilize watershed boundaries to manage activities (Conservation Ontario, 2010; Robins, 2007). In 2002, the Nova Scotia Provincial Drinking Water Strategy was published to address management of water resources from a human consumption standpoint, and identified a need to manage watershed resources beyond drinking water (2002). In 2007, the Environmental Goals and Sustainable Prosperity Act was created to achieve a more environmentally and economically sustainable province by 2020. This act outlined the province's goal to develop a water strategy to manage competing demands for water resources (Nova Scotia, 2008). In 2010, the Water for Life water strategy was published (Nova Scotia, 2010), which outlined a broad set of integrated water management goals for the province, including:

- sharing resources and information while working with partners to manage water
- renewing policies as appropriate to enhance capacity for integrated water management
- establishing a Nova Scotia Water Advisory Group to work in partnership with government and advise it on the implementation of integrated water management and the strategy.

The Water for Life strategy was initially viewed as a valuable tool for improving water resource management in the province, but according to Jocelyn Rankin, the former chair of the Nova Scotia Environmental Network's Water Caucus, there was criticism about its effectiveness due to its lack of clear targets and timeframes (personal communication, 25 February 2016). The strategy led to the creation of valuable water resource management initiatives throughout the province, but many of these did not last long enough to adequately address the issues they were targeting.

Unlike Ontario or Manitoba, Nova Scotia does not have formalized units in which to manage water resources on a watershed scale. Instead, there are several community groups and non-profit organizations throughout the province that operate at the local watershed level and vary considerably in their make-up and the issues they target. These groups have made strides in improving the consistency and quality of information being collected, but integration of this knowledge with management at the municipal and provincial levels still remains limited.

CARP: a Nova Scotian case study in integrated watershed management

Organizational structure and operation

CARP is an environmental non-profit organization and registered charity incorporated in 1990 with a mission to enhance the ecological health of the Annapolis River watershed through science, leadership and community engagement (Acadia Centre for Social and Business Entrepreneurship [ACSBE], 2012). CARP is based in Annapolis Royal, Nova Scotia, and conducts the majority of its work within the Annapolis River watershed.

A 15-member volunteer board of directors that represents the interests of various stakeholder groups and watershed residents governs the organization. CARP employs an executive director, who oversees the operational aspects of the organization and answers to the board of directors; an administrative manager, who is responsible for human resources and financial management; and a programme manager, who is responsible for managing the delivery of CARP's various programmes and projects. The group employs numerous project leaders and staff on a contractual basis, and relies on the efforts of community volunteers and partners to conduct its activities.

Funding for CARP comes primarily from grants obtained through applications for project-specific funding from a variety of sources, including approximately 31% from the government of Canada, 26% from the province of Nova Scotia, 42% from corporate contributions, foundations and NGOs, and 1% from private donations. Annual expenditures since 2010 have been in the range of CAD 550,000–650,000.

The Annapolis River watershed

The Annapolis River watershed (Figure 1), which is part of the Gulf of Maine watershed, is the third-largest in Nova Scotia, covering portions of Kings, Annapolis and Digby Counties. It has an approximate area of 2250 km^2 and includes all of the lakes, streams and wetlands that drain into the Annapolis River and downstream to the Annapolis Basin, where it meets the Bay of Fundy (Sutherland, 2003).

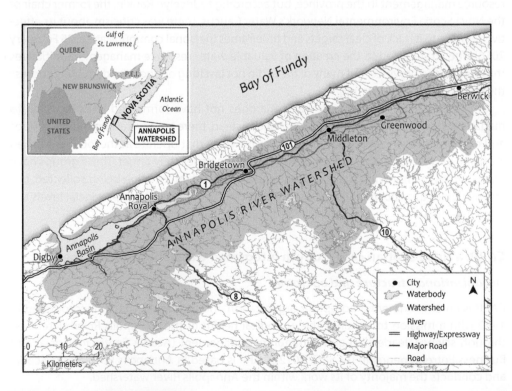

Figure 1. Annapolis River Watershed.

96

The Annapolis River watershed forms part of the Annapolis Valley, an important agricultural region (Timmer, de Loë, & Kreutzwiser, 2007). The region provided its Mi'kmaq inhabitants with prosperous livelihoods for thousands of years prior to settlement by Europeans in the early 1600s (Sutherland, 2003). Since that time, water quality has suffered due to land-based impacts of agriculture, municipal sewage treatment facilities, malfunctioning on-site septic systems, uncontrolled livestock access to waterways, industrial effluents, and runoff from urban areas (Timmer et al. 2007).

History

Annapolis Valley Affiliated Boards of Trade: application for heritage river status

The history of CARP is rooted in the actions of the Annapolis Valley Affiliated Boards of Trade, which in 1985 formed the Annapolis River Task Force, comprising a broad cross-section of representatives from the local business, cultural and government sectors. The group was tasked with preparing background information in an effort to have the Annapolis River become the first river designated under the Canadian Heritage River System (Clean Annapolis River Project [CARP], 1993). The Canadian Heritage River System was established in 1984 by federal, provincial and territorial governments to conserve rivers with outstanding natural, cultural and recreational heritage, to give them national recognition and to encourage the public to enjoy and appreciate them (http://chrs.ca/about/).

Diane LeGard, past member of the Annapolis Valley Affiliated Boards of Trade and the Annapolis River Task Force, stated that it was felt that the Annapolis River was a strong candidate for designation based on its history as the earliest existing European settlement north of St. Augustine, Florida, having been settled in 1604. There was a strong desire to highlight this history and heritage in hopes of promoting tourism to the region (personal communication, 8 December 2015).

According to Stephen Hawboldt, founding member of CARP and past executive director, the task force put together a rationale that was primarily focused on the cultural history of the area. The application was rejected due to a combination of the ecological degradation the river had suffered through a long history of anthropogenic impacts, and the focus of the Canadian Heritage River System Technical Planning Committee on natural history at the time the application was made (personal communication, 5 February 2015).

The rejection drew immediate attention from community leaders to the environmental concerns facing the river. This led to an expansion of the Annapolis River Task Force's mandate to include exploring and seeking solutions to these issues. The public was engaged in the process through a series of forums that resulted in heightened awareness of local environmental issues among watershed residents (CARP, 1993).

Formation of the Clean Annapolis River Project

Simultaneous with the Annapolis Valley Affiliated Boards of Trade's efforts to have the Annapolis River recognized as a heritage river, a group of private- and public-sector scientists were becoming more concerned about the state and treatment of estuaries in Atlantic Canada. This led to the formation of an informal group called the Atlantic Estuaries Cooperative Venture (CARP, 1993). According to Hawboldt, the group held the belief that the best approach to addressing human impacts on the estuarine environment was to develop community-based solutions informed by good science, and so they proposed the establishment of a pilot site to demonstrate this approach in each of the Maritime Provinces (personal communication, 5 February 2015).

In April 1998, two Atlantic Estuaries Cooperative Venture associates, Dr Graham Daborn of the Acadia Centre for Estuarine Research and Joe Arbor of Environment Canada, approached the Annapolis River Task Force to request their participation in the venture. A subcommittee was formed to consider the proposal and recommend next steps, which led to a public meeting in 1990, where the decision was made to form CARP (1993). Hawboldt recalls that the intent was to establish an organization that used a positive, collaborative approach to engage multiple stakeholders in the process of identifying high-priority environmental issues and community-based solutions, guided by the best science available (personal communication, 5 February 2015).

The Atlantic Coastal Action Program

As a component of the government of Canada's 1990 Green Plan, the Atlantic Coastal Action Program (ACAP) was developed in 1991 as a six-year initiative to restore Atlantic Canada's coastal areas in order to protect the livelihoods and communities of coastal residents (Environment Canada, 1995a). The intent of the programme was to build capacity in communities to address environmental challenges using a science-based, regional ecosystem approach, engaging all sectors of the community in the identification of issues and solutions. Environment Canada (1995b) espoused five primary guiding principles for ecosystem initiatives, including ACAP, which may be outlined as:

(1)   Ecosystem approach. It is important to consider all of the elements that make up an ecosystem and to examine how they inter-relate.

(2)   Partnerships. To be truly effective, an ecosystem management plan should be designed and implemented by those who are affected by and have an interest in the outcome.

(3)   Environmental citizenship. Canadians should be provided with timely, accurate and understandable information to help them make informed decisions and promote action.

(4)   Science. The information used in decision making should go beyond the pure sciences and encompass the broadest range of disciplines.

(5)   Environment Canada should provide leadership in promoting ecosystem sustainability.

According to Hawboldt, ACAP grew out of the efforts of the Atlantic Estuaries Cooperative Venture because of the established connections CARP had with the organization. CARP became the first group to sign a contract with ACAP, in the fall of 1991 (personal communication, 5 February 2015). In his article on ACAP, Robinson (1997, p. 130) describes CARP's integration into the programme:

> CARP was incorporated into ACAP as it espoused many of the same aims and ideals as those being promoted by Environment Canada through the ACAP scheme, and a strong set of links had already been established between CARP and other environmental organizations throughout the Bay of Fundy and Gulf of Maine areas.

The programme grew to cover 14 sites distributed throughout Atlantic Canada, and provided funding and support that helped build capacity at each of these sites to address local environmental concerns prioritized by the community. Each organization was provided CAD

50,000 per year for up to five years to hire a coordinator and maintain an office, as well as varying amounts of seed project funding to provide leverage for additional resources from other funding sources and community partners. The investment in the programme through the federal Green Plan for 1991–1996 was CAD 8.9 million (Environment Canada, 1995a). The success of the programme led to its continuation beyond the original six years, and an additional CAD 13,528,500 was invested in the programme from 1997 to 2001 (Gardner Pinfold Consulting Economists Limited, 2002).

Beginning in 2009, ACAP began transitioning into the Atlantic Ecosystem Initiative, gradually shifting focus from a regional capacity-building approach to a larger-scale ecosystem approach, and adopting a project-based funding structure. Beginning in 2015, the programme would seek to fund larger projects encompassing the entire Atlantic region, favouring partnerships and coalitions among groups, with the goal of increasing its effectiveness in improving the health, productivity and long-term sustainability of the Atlantic ecosystem. Eligibility to compete for Atlantic Ecosystem Initiative funds expanded beyond the established sites to include all NGOs, coalitions and networks, research and academic institutions, and Aboriginal governments and organizations (Mercer, personal communication, 18 September 2014). The overall shift in the focus and funding structure of these programmes has led CARP to seek alternative means of maintaining the financial and human capacity to address local environmental concerns by developing stronger partnerships with like-minded organizations and seeking to become more independent of the support that Atlantic Ecosystem Initiative traditionally provided.

Integrated planning approach for the watershed

Since its beginning, CARP has taken an approach to watershed-scale environmental management that integrates and supports the efforts of multiple stakeholders and decision makers. Hawboldt describes the importance of organizational composition to the adoption of an integrated management strategy: "CARP was formed with the involvement from a unique combination of representatives from the private sector, academia, various levels of government, and local citizens. This allowed the organization to become well known and accepted by multiple sectors very quickly" (personal communication, 4 February 2015).

Defining an organizational strategy that emphasizes an integrated approach to managing the Annapolis River watershed has been supported throughout CARP's existence by ongoing planning processes that have charted CARP's direction. Beginning in its first phase, ACAP required that participating groups develop a comprehensive environmental management plan, the goal of which was to assess the environmental quality of each ACAP site's geographic area of focus, identify the sources of environmental problems, and determine the remedial and conservation measures (Environment Canada, 1995b). CARP's first environmental management plan, titled "Our Watershed, Our Responsibility", was developed through consultations with a focus group comprising various representative sectors, including environment, health, business and recreation. The draft document was revised based on feedback provided by 22 stakeholder groups, including municipal, provincial, federal and First Nations governments, academia, the non-profit sector and business (Garret, 1996). The document identified numerous actions under the categories of water, air and land. The specific objectives for water included improving water quality using a safe recreational limit of 200 colony-forming units of faecal coliform per 100 mL as a target, ensuring no net loss of aquatic habitat, promoting the integration of various water uses such as the aquaculture industry and the commercial fisheries, ensuring the long-term economic sustainability of fisheries, and ensuring that low-flow requirements were maintained at levels sufficient to support aquatic life (Garret, 1996). This guided CARP's activities from 1996 to 2003.

An updated plan, titled "Moving Forward: An Environmental Management Plan for the Annapolis Watershed", was created in 2003. This plan used a similar approach of stakeholder engagement through a series of open houses, and surveys to capture input from community members on their perspectives regarding environmental issues in the watershed. Two organizations were engaged to represent targeted sectors through focus groups. The Western Valley Development Authority was asked to list, prioritize and suggest solutions to local environmental problems through completion of a survey, while the Annapolis Basin Working Group addressed fisheries-related issues and the environmental state of the Annapolis Basin through a facilitated discussion (Sutherland, 2003). The information was synthesized into a prioritized list of environmental issues in the Annapolis watershed, with recommended solutions. A list of broad water-related goals was presented under separate categories of water quality, water quantity and habitat. The water quality goals included setting more stringent targets for faecal coliform levels than in the 2003 document, raising community awareness of water quality issues, and identifying and eliminating surface water and groundwater pollutants and their sources. The water quantity goals included maintaining an adequate water supply for all current and future users, and establishing water conservation practices among all users in the watershed. Goals relating to aquatic habitats included establishing and maintaining habitat to preserve species diversity and healthy populations, and discouraging the introduction of invasive species to preserve native species and their habitats. The document was reviewed by CARP's board of directors, as well as a project advisory committee, prior to acceptance of the environmental plan for the Annapolis watershed (Sutherland, 2003). The process of creating the environmental plan demonstrated an evolution in integrated watershed management that was shaped by 13 years of experiences and lessons. In 2003, the integrated watershed management process included a more structured engagement process that utilized surveys to capture stakeholder information in an organized fashion, in addition to the discussions held at open houses and focus group sessions.

In 2008, CARP retained the services of the Acadia Centre for Social and Business Entrepreneurship (ACSBE) to facilitate a six-month strategic planning process aimed at reviewing organizational vision and direction in order to ensure CARP's continued relevance to its constituents, staff, board and stakeholders (ACSBE, 2009). Background information to inform the process was gathered through online surveys of CARP staff, board members and stakeholders. Survey questions aimed to gather participants' perspectives on organizational purpose, strengths, weaknesses, opportunities, potential partnerships, and challenges. The results informed the design of the planning process, which took place over four sessions between January and April 2009 (ACSBE, 2009). The outcome was a strategic plan that differed from previous environmental management plans in that it defined broader goals and specific objectives relating directly to CARP's mission statement, rather than a specific list of environmental concerns and solutions (Table 1).

The 2009 plan was updated in 2012 as a required component of an Atlantic Ecosystem Initiative contract. The process, once again facilitated by ACSBE, included board members, staff and stakeholder surveys that informed a series of three reflective sessions attended by CARP staff and board members (ACSBE, 2012). CARP's vision, mission, goals and objectives

Table 1. The Clean Annapolis River Project's strategic goals.

1	To continually assess the ecological health and environmental stressors of the Annapolis River watershed
2	To identify and establish priorities and projects to enhance the ecological integrity of the Annapolis River watershed
3	To empower all decision-makers with the knowledge, tools, and research to make ecologically sound decisions with respect to the Annapolis River watershed
4	To facilitate collaboration among all stakeholders to address the environmental challenges facing the Annapolis River watershed
5	To engage local community members in the restoration and protection of the Annapolis River watershed in ways that are meaningful, relevant, and fun

were reviewed. The five strategic goals established in 2009 were retained, as they were thought to strongly reflect CARP's integrated approach to watershed management.

Integrated watershed management in action

Guided by the principles outlined in the organization's mission and strategic goals, CARP seeks to undertake projects that contribute to key parts of an integrated management strategy for the Annapolis River watershed by collecting high-quality science, providing leadership and fostering community engagement. These principles, as well as the strategic goals outlined in Table 1, provide the guiding framework for the work undertaken by the organization. Projects often seek to collect and disseminate environmental data and incorporate input from the public and stakeholders. These data are useful in guiding internal planning processes, as well as for informing stakeholders and decision makers such that they are able to take ecologically responsible actions. CARP also strives to engage community members and landowners in many of its projects as part of its efforts to improve stewardship in the greater watershed and to enhance and improve ecosystem health.

Community engagement: fostering stewardship

As a community-based watershed group, CARP is in a unique position to facilitate collaboration with multiple stakeholder groups, academia, government and local communities in order to engage community members, foster stewardship and improve watershed management. CARP works with local communities in the Annapolis River watershed on projects focused on enhancing environmental stewardship, such as those it undertakes with the agricultural community. Agricultural activities are one of the predominant land uses in the watershed (Brylinksy, 1992; Sutherland, 2003). CARP has been working for several years with farmers to deliver programmes targeted at reducing the impacts of farming activities on the Annapolis River ecosystem. Partnerships with other agencies, such as provincial government departments and non-governmental groups (e.g. Ducks Unlimited), have provided opportunities to share resources and knowledge more effectively in the delivery of CARP's agricultural stewardship initiatives. Projects to date have been aimed at reducing the amount of contaminants entering waterways from agricultural runoff, improving the amount and quality of riparian buffers along shorelines, fencing cattle away from waterways, and creating farm stewardship agreements with local farmers and landowners. Some agricultural producers, although ecologically minded, can be reluctant to adopt new farm practices aimed at improving ecological health, as they often require significant time, energy and resources. CARP helps surmount these barriers by consulting with farmers on the development of solutions that satisfy mutual needs, and on the implementation of actions identified through a collaborative process. CARP is often able to contribute materials and human resources to efforts to control erosion, improve cattle management, manage nutrients and re-establish degraded riparian buffers, on the condition that farmers commit to ongoing stewardship by following established best management practices. By providing targeted outreach to the agricultural community through presentations to youth and industry groups and through attendance at public events, CARP also seeks to educate local farmers and the public on the importance of adopting agricultural stewardship practices.

Providing leadership: informing and influencing decision making

Although many community-based environmental stewardship groups have not had the opportunity to play a significant role in watershed management (Sharpe & Conrad, 2006), strides have been made in recent years towards building partnerships and influencing decision making at both local and provincial scales. CARP strives to be a leader at both the local level and the provincial scale in the mentorship it provides to volunteers in collecting high-quality scientific data and in leading programmes that can be adopted by other community groups in their own watersheds to help improve decision making on a larger scale. CARP's water quality monitoring programmes, for example, provide long-term baseline data on surface water quality in the Annapolis River watershed and monitor physical, chemical and biological water quality parameters, such as water temperature, pH, dissolved oxygen, conductivity, turbidity, benthic macro-invertebrates, nutrients and total dissolved solids. These programmes, as well as other water quality initiatives in which CARP participates with its partners, provide valuable ecological monitoring information that can be used by decision makers and community partners to determine what actions are needed to manage watershed health.

CARP's water quality monitoring initiatives began in the early 1990s with the initiation of a long-term, volunteer-based water quality monitoring programme, designed and developed in collaboration with several partners, including the Acadia Centre for Estuarine Research, several departments within the provincial government, and Nova Scotia Community College (Beveridge, Sharpe, & Sullivan, 2006; Freeman, 2013; 1995). In the early 1990s, it was recognized that to achieve consistent, affordable long-term monitoring data and information on water quality trends it would be necessary to promote greater public participation in the data collection and remove some of the fiscal pressure from regulatory agencies (Brylinksy, 1992). This led to the development of water quality monitoring programmes that utilized volunteers in the collection of scientific data, which were developed in consultation with multiple partners. CARP's longest-running water quality monitoring programme, Annapolis River Guardians, has collected 24 years' worth of water quality data through the use of volunteer river guardians. Over the years, the number of volunteers participating in the programme has ranged from 8 to 43; they collect data on parameters such as water temperature, dissolved oxygen, *Escherichia coli* and weather conditions. CARP staff collect additional parameters such as pH, total dissolved solids and turbidity, and transport volunteers' samples to a local laboratory for analysis. CARP's water quality monitoring programmes are continually being evaluated and adapted to improve the quality and relevance of the data collected.

Danielsen et al. (2014), in a multi-country assessment, found no significant differences in data gathered from scientist and volunteer-based data collection regimes. Similarly, Shelton

(2013) studied the accuracy of water quality data collected by volunteers as compared to by scientists, and found that for most water quality parameters studied, similar results were obtained. CARP's core water monitoring programmes utilize a combination of scientists and community volunteers to collect high-quality scientific data, and encourage local steward-ship through the public participation, education and volunteerism involved in monitoring local waterways. Involving the local community also provides CARP staff with valuable anec-dotal knowledge and information from members who have lived in the watershed for many years. Such information can often help guide local management and watershed planning initiatives.

CARP ensures that accurate volunteer data are collected using thorough volunteer train-ing programmes, as well as comprehensive quality assurance and control protocols. The water quality information collected by and for CARP is made publicly available and is used by CARP to guide decision making on future actions and monitoring needed in the water-shed. CARP also collaborates with several partners in furthering the development of water quality monitoring programmes in other regions of Nova Scotia, and in providing data for multi-agency use. For example, as part of CARP's annual water quality monitoring pro-grammes, several benthic macro-invertebrate samples are collected using procedures devel-oped for Environment Canada's Canadian Aquatic Biomonitoring Network, a national biomonitoring programme (Environment Canada, 2012). This allows the collection of con-sistent and comparable data nationally, which is shared with Environment Canada for their own use, and for addition to their online databank.

As an environmental NGO with long-term community-based monitoring experience, CARP has been in a position to support groups in Nova Scotia and other parts of Atlantic Canada in the development of their water quality monitoring programmes. In recent years, CARP has had the opportunity to partner with the Community Based Environmental Monitoring Network at Saint Mary's University on their CURA H_2O programme. The goal of the programme is to increase community capacity for integrated water monitoring and management in Canada and internationally (http://cbemn.ca/cura-h2o-archives/about/). The CURA H_2O programme has resulted in great strides towards that end by developing resource and training materials for community-based organizations, creating and distribut-ing standardized equipment kits to programme partners, running an online water quality monitoring certification programme, developing an online database to host and display water quality monitoring data collected by programme partners, and leading research on community-based monitoring to inform the evolution of its practice. The Community Based Environmental Monitoring Network is conducting ongoing efforts towards expanding the effective practice of integrated water monitoring and management, in which CARP will continue to play a supporting role.

Collecting high-quality science: integrated management planning for sub-watersheds
The majority of CARP's projects are focused on the collection of sound scientific data, which is most effective when collected in a considered, targeted manner. The work that CARP has been undertaking with its fish habitat and aquatic connectivity monitoring projects provides an example, as they are part of an initiative to develop detailed sub-watershed management plans in the Annapolis River watershed to guide data collection and restoration actions. CARP's aquatic connectivity work began in 2007, after it was recognized that habitat frag-mentation was an issue that needed to be assessed in the watershed (Hicks, 2007). Over the

course of five years, CARP field staff visited almost 90% of the road-watercourse crossings in the Annapolis River watershed and completed detailed assessments on approximately 25% (Freeman, 2014a). Barrier crossings that have been assessed are prioritized for remediation. In the early years of the programme, CARP assessed watercourses in the watershed that were close to the main stem of the Annapolis River. However, as the Annapolis River watershed comprises many large sub-watersheds, this approach was a challenging and unfocused way to target areas for restoration. In 2012, a sub-watershed approach was adopted to assess connectivity issues within one entire sub-watershed at a time in order to improve watershed planning initiatives (Freeman, 2014b; Freeman, 2013; Wagner, 2013).

Several sub-watersheds in the Annapolis Valley were identified as warranting high priority in restoration work, due to their natural water quality characteristics (Dickinson, 1992; Parker, Halliday, & Pearle, 1994; Wagner, 2013). These watersheds have been targeted for the creation of management plans to guide future restoration actions in the watershed. Restoration needs are evaluated and prioritized on a sub-watershed scale, and management plans are developed which incorporate knowledge and insight acquired from local communities and partners. Because of its non-governmental nature, CARP is able to develop much closer working relationships with members of the local community, such as anglers, hunters and other recreational users of the river. This enables the organization to gain valuable information on the local watershed using tools such as angler surveys, workshops and meetings, which assists with the development of sub-watershed management plans, and allows more extensive public input into how the watershed is managed.

Habitat connectivity and quality are assessed at a local scale through sub-watershed planning processes, and protocols have been developed that are continually being adapted and improved through consultation with partnering organizations such as the Nova Scotia Salmon Association's NSLC Adopt A Stream programme and Fisheries and Oceans Canada. Likewise, the sub-watershed management plans developed by CARP staff are working documents that are continuously updated as restoration work is completed, and new management strategies are adapted to changing environmental and socio-economic conditions.

Successes and challenges

An integrated watershed management approach has a number of benefits but also many challenges that need to be overcome in order to achieve healthy aquatic ecosystems. Some of the successes and challenges of watershed management in Nova Scotia in a community-based framework have been documented (Conrad & Hilchey, 2011; Sharpe & Conrad, 2006). As a community-based organization that has been monitoring and developing watershed management plans for the Annapolis River watershed for over 20 years, CARP has more established methods of operation than many other community-based monitoring groups across Nova Scotia. Its capacity to successfully operate, however, is largely influenced by the same financial, human, institutional, political and social elements that similarly affect other provincial community-based organizations.

Institutional capacity
The partnerships that CARP has developed over the years with academia, government, the non-profit sector and other community-based monitoring groups have been instrumental in the provision of scientific expertise and resources for the development and execution of

many of CARP's projects. These sorts of partnerships help enhance the institutional capacity of local groups, such as CARP, to deliver programmes in a more cost-effective manner than if the same work were carried out solely through governmental departments (Durley, de Loë, & Kreutzwiser, 2003; Timmer et al., 2007). CARP's ability to undertake technical and scientific programmes is enhanced because of the expertise provided by its partners in both conceptualization and implementation. Use of volunteers in data collection helps improve CARP's capacity to collect monitoring data and to implement projects, and works well when a scientifically trained project leader oversees the volunteers in the collection of data. Data inaccuracy and fragmentation challenges have been cited as one of the reasons that data collected by community-based groups is often disregarded by decision makers (Conrad & Hilchey, 2011; Conrad & Daoust, 2008; Sharpe & Conrad, 2006). However, CARP's use of quality assurance and quality control procedures in its data collection methods, its adherence to accepted sampling procedures, and the planning it undertakes to determine the viability of projects help address and overcome many of these challenges, and allow the organization to successfully and consistently collect reliable data.

Integration of data with those from other partners has been improving over the past several years, but remains limited. There is a lack of consistency between data-sets, making comparability of data challenging (Sharpe & Conrad, 2006). In recent years, however, strides have been made by CURA H_2O to address data-collection inconsistencies by building the institutional capacity of many community-based groups to collect comparable, accurate data. Improvements still are needed, however, to integrate the monitoring data being collected by community-based groups with governmental data-sets, to better inform and guide decision making related to watershed management.

Financial and human capacity

CARP's ability to successfully address watershed health is largely influenced by elements of financial and human capacity. Projects have been successfully developed to address many issues identified in the watershed, such as water quality, riparian land use and fish habitat fragmentation. One of the foremost challenges in strengthening existing projects or exploring new programmes, however, is the lack of adequate financial and human resources to support the organization's operational and technical needs. This is a common and recurring issue facing small communities and other community-based organizations alike (Conrad & Hilchey, 2011; Dugas, 2009; Sharpe & Conrad, 2006; Timmer et al. 2007; Whitelaw, Vaughan, Craig, & Atkinson, 2003). In CARP's experience, the types of funding sources available make watershed planning and management challenging for many community-based groups, as few funders are willing to fund planning initiatives, preferring to finance direct, observable actions. This can result in projects lacking adequate planning and preparation in determining what the most appropriate strategy is for moving forward. CARP has developed management plans for some of the high-priority sub-watersheds in the Annapolis River watershed, which provide a strategic, consistent, clear path forward in identifying, recognizing and addressing sub-watershed management needs. Development of these plans requires time, money and human resources, which are often in short supply.

The financial capacity of community-based organizations like CARP to support monitoring and restoration projects can be limited by the process of securing funds, as it requires a certain amount of technical skill and human resources to develop applications for funding, and financing the overhead required to secure funds can be challenging. The discontinuous and fluid nature of funding opportunities also often restricts what types of projects can be funded on a yearly basis, which makes continuous long-term funding of monitoring projects challenging.

The fragmented nature and uncertainty about sources of funding also means that there is no reliable source of core funding for an organization such as CARP, which has an adverse effect when it comes to building and maintaining an adequate human resource capacity in a small community (Timmer et al. 2007). It can be difficult to attract and retain skilled staff due to the variable nature of funding opportunities, limited resources for professional development, and lack of stable employment opportunities. The funding challenges and small rural community structure of the Annapolis River watershed make recruitment and maintenance of skilled and experienced personnel challenging, often resulting in high staff turnover rates, which can create additional institutional challenges in ensuring project continuity and consistency. CARP has core staff who are able to be retained on a yearly basis, which improves its ability to develop some skilled professionals, but extra funds for skill development opportunities are limited. The situation also does not reflect the human capacity of other community-based groups in Nova Scotia, whose organizational profiles and human resource capacity vary considerably (Sharpe & Sullivan, 2003).

While it can be a challenge to recruit and maintain skilled professionals in the community-based non-profit world, it can be similarly challenging to find dedicated long-term volunteers to commit to monitoring programmes. CARP has been successful in recruiting some committed individuals, but volunteer turnover remains high, and resources must be dedicated to advertising and to training new volunteers on an annual basis. Acknowledgement of volunteers' contributions and efforts is essential when it comes to maintaining committed volunteers. Showing volunteers that the information they collect is valued and useful helps uphold volunteer enthusiasm for projects. Aligning volunteer schedules and abilities to project needs and work hours can sometimes be difficult. Often the volunteers most able to commit to daytime hours are retirees, and while these individuals may be able to contribute substantial amounts of time, they are not always able to meet the physical demands of some aspects of field data collection. On the other hand, those volunteers who are generally quite able to meet the demands of more physically challenging projects often have careers of their own and are much more limited in the amount of time they can commit and when they are able to do so.

Political capacity

Managing the Annapolis River on a watershed scale is also challenging because the watershed spans multiple municipal and county boundaries and comprises numerous communities governed by various town and municipal councils with divergent political agendas and priorities. This is a problem many watershed groups face: watershed boundaries that do not coincide with political ones (Blomquist & Schlager, 2005; Timmer et al. 2007). It adds layers of complexity to the successful management of a watershed when there are multiple agencies involved and a lack of communication between them. Local community groups in Nova Scotia such as CARP have the political capacity to carry out work that crosses political boundaries, but do not have the governance ability to manage or regulate various land use activities within a watershed. Therefore, increased collaboration and integration of monitoring and governance are necessary to improve watershed level management. CARP has made good progress in building links with the governing bodies in the Annapolis River watershed, though further efforts to strengthen these relationships are necessary. Improved cooperation and communication among various levels of government, stakeholders, water users and community groups is essential to maximize efficiency and minimize duplication of data collection, results sharing and decision making. This is not a new concept, and has been a recurring theme throughout past analyses of challenges in water resource management in Nova Scotia (Conrad & Hilchey, 2011; Sharpe & Conrad, 2006; Timmer et al. 2007). The process of streamlining efforts has begun through the development of online data-sharing databases for use of community-based groups across Nova Scotia and the Atlantic Provinces, and in the standardization of protocols used across select levels of government and NGOs. This work should continue further in order to create closer links and working relationships among governmental departments and other water users and stakeholders, and to foster greater consistency in data collection and interpretation in the province.

Obtaining the cooperation of some stakeholder groups can be a challenge when government is involved (or perceived to be involved). Although CARP is not a regulatory agency, many private landowners mistrust organizations with an environmental mandate. There is a fear that community-based monitoring organizations will act as whistle blowers, and many landowners and resource users are reluctant to collaborate on ecological assessment and improvement projects for that reason.

Social capacity

The social element of watershed management is an important consideration when it comes to addressing issues affecting watershed health. As a non-regulatory body, it is important for CARP to secure public support and participation in order to effectively deliver programmes that address watershed issues. Operating from a non-governmental standpoint allows CARP to develop closer ties to the communities within the Annapolis River watershed, by allowing CARP to take a collaborative approach with landowners rather than a regulatory one. Since 2003, CARP has worked with over 68 landowners to improve stewardship practices and enhance riparian health throughout the Annapolis River watershed. It can be challenging, however, to overcome the apathy of many property owners with regard to improving the ecological health of their properties. When it comes to adopting land stewardship practices for the sole sake of watershed health, relatively few property owners are willing to cooperate, which limits where effective action can occur. Many, though not all, require a sense of direct economic or personal benefit to motivate them to collaborate on stewardship initiatives.

Just as it can be a challenge to bring landowners on board with stewardship and watershed management initiatives, it can also be difficult to reach out to the general public. Although CARP has close links to the local community through its work, considerable resources are still required to deliver outreach to the public and build social capital (Conrad & Daoust, 2008). Substantial time and effort are required to develop outreach and education programmes, create information pamphlets, deliver events and so on in an attempt to raise general awareness of watershed management issues, and to recruit volunteers. These activities are difficult to fund but are necessary to the process of building community links and partnerships, and fostering a sense of community stewardship.

Gaining public knowledge and input can still sometimes prove challenging, as there is often a dichotomy between those who want to provide input and those who actually do so when given the opportunity. CARP has made progress in gaining greater public input through use of tools such as meetings, workshops, presentations, public events, surveys and interviews. Such outreach requires the dedication of valuable resources that many groups may not have the capacity to dedicate or acquire for those tasks (Timmer et al. 2007). Engaging the public through the use of volunteers for scientific data collection has also been valuable in both reducing the human resource cost of using staff for data collection, and improving the awareness and general scientific competence of the members of the public who assist in data collection. In CARP's experience, engaging the public in projects has had a twofold impact: it has enhanced education on local and universal environmental impacts in the watershed; and it has helped address citizen apathy towards environmental matters by engaging volunteers in learning first-hand about watershed issues through field participation. Volunteers for CARP generally demonstrate a much keener interest in watershed management issues when they are given the opportunity to observe, learn and participate in monitoring efforts in their own community. Similar observations have been made by others in the community-based monitoring field, where it has been noted that community-based watershed management provides a local component which encourages greater vested interest and citizen commitment to long-term issues (Conrad & Hilchey, 2011; Dugas & Sharpe, 2009).

While many challenges exist in the realm of community-based watershed management, they are not insurmountable. The benefits of more cost-effective monitoring, increased public input and involvement in monitoring, strengthened community stewardship and improved ecological health far outweigh the challenges.

Conclusion

The approach that CARP takes today in managing water resources in the Annapolis River watershed has largely been defined and influenced throughout the organization's history and development. CARP's community-based approach to watershed management provides a means of monitoring and evaluating ecosystem health locally. The strategic goals of the organization guide the management of water quality on an ecosystem scale and promote an integrated approach to addressing ecological issues in the watershed. While there are still many challenges in the sphere of community-based watershed management, overcoming these hurdles is necessary in order to improve local ecosystem health. In light of the growing difficulties facing water resource monitoring and management in Canada, it is not only beneficial but increasingly necessary to develop watershed management approaches that foster improved collaboration, communication and informed decision making in order to continue to adequately manage and safeguard aquatic systems in Canada for future generations.

Disclosure statement

No potential conflict of interest was reported by the authors.

References

Acadia Centre for Social and Business Entrepreneurship. (2009). *Clean Annapolis river project: Strategic planning 1999*. Wolfville, NS: Acadia Centre for Social and Business Entrepreneurship.

Acadia Centre for Social and Business Entrepreneurship. (2012). *Clean Annapolis river project: Strategic planning*. Wolfville, NS: Acadia Centre for Social and Business Entrepreneurship.

Beveridge, M. A., Sharpe, A., & Sullivan, D. (2006). *Annapolis river 2005 annual water quality monitoring report*. Annapolis Royal, NS: Clean Annapolis River Project.

Blomquist, W., & Schlager, E. (2005). Political pitfalls of integrated watershed management. *Society and Natural Resources, 18*, 101–117. doi:10.1080/08941920590894435

Brylinksy, M. (1992). *Procedures manual for the clean annapolis river project river guardian programme* (ACER Publication Series No. 24). Wolfville, Nova Scotia: Acadia Centre for Estuarine Research, Acadia University.

Clean Annapolis River Project. (1993). *Multistakeholder summary: Clean Annapolis river project*. Clementsport, NS: Clean Annapolis River Project.

Conrad, C. T., & Daoust, T. (2008). Community-based monitoring frameworks: Increasing the effectiveness of environmental stewardship. *Environmental Management, 41*, 358–366. doi:10.1007/s00267-007-9042-x

Conrad, C. C., & Hilchey, K. G. (2011). A review of citizen science and community-based environmental monitoring: Issues and opportunities. *Environmental Monitoring and Assessment, 176*, 273–291. doi:10.1007/s10661-010-1582-5

Conservation Ontario. (2010). *Integrated watershed management: Navigating Ontario's future*. Newmarket, ON: Conservation Ontario.

Danielsen, F., Jensen, P. M., Burgess, N. D., Altamirano, R. A., Alviola, P. A., Andrianandrasana, H., … Young, R. (2014). A Multicountry assessment of tropical resource monitoring by local communities. *BioScience, 64*, 236–251. doi:10.1093/biosci/biu001

Department of Environment and Labour. (2002). A drinking water strategy for Nova Scotia – A comprehensive approach to the management of drinking water. Province of Nova Scotia, 42 pp.

Dickinson, A. (1992). *Initial study of pH values of streams within the Annapolis River watershed related to Torbrook Formation*. Annapolis Royal, NS: Clean Annapolis River Project.

Dugas, K. (2009). *Discussion paper on watershed management strategies for Nova Scotia, with a focus on the Annapolis river watershed*. Annapolis Royal, NS: Clean Annapolis River Project.

Dugas, K., & Sharpe, A. (2009). Wading. In *Watershed management in Nova Scotia*. Report of workshop outcomes, March 26 & 27, 2009. Annapolis Royal, NS: Clean Annapolis River Project.

Durley, J. L., de Loë, R. C., & Kreutzwiser, R. D. (2003). Drought contingency planning and implementation at the local level in Ontario. *Canadian Water Resources Journal, 28*, 21–52.

Environment Canada. (1995a). *Ecosystem initiatives in environment Canada: A synopsis*. Ottawa, ON: Environment Canada.

Environment Canada. (1995b). *Guiding principles for ecosystem initiatives*. Ottawa, ON: Environment Canada.

Environment Canada. (2012). *Canadian aquatic biomonitoring network field manual: Wadeable streams*. Dartmouth, NS: Environment Canada.

Freeman, L. (2013). *Annapolis river 2012 annual water quality monitoring report*. Annapolis Royal, NS: Clean Annapolis River Project.

Freeman, L. (2014a). *Broken Brooks 2014: Improving In-stream fish habitats through restoration*. Annapolis Royal, NS: Clean Annapolis River project.

Freeman, L. (2014b). *Nictaux river sub-watershed management plan*. Annapolis Royal, NS: Clean Annapolis River Project.

Gardner Pinfold Consulting Economists Limited. (2002). *An evaluation of the Atlantic Coastal Action program: Economic impact and return on investment*. Halifax, NS: Gardner Pinfold Consulting Economists Limited.

Garret, J. A. (1996). *Our watershed, our responsibility: Annapolis environmental management handbook*. Annapolis Royal, NS: Clean Annapolis River Project.

Hicks, K. (2007). *Broken brooks: 2007 project report*. Annapolis Royal, NS: Clean Annapolis River Project.

Nova Scotia. (2008). *Environmental goals and sustainable prosperity act: Annual progress report 2008*. Halifax, NS: Government of Nova Scotia.

Nova Scotia. (2010). *Water for life: Nova Scotia's water resource management strategy*. Halifax, NS: Government of Nova Scotia.

Parker, M. A., Halliday, C. A., & Pearle, M. J. (1994). *Fish habitat restoration and training project*. Annapolis Royal, NS: Clean Annapolis River Project.

Proszynska, G. (1995). *Annapolis river guardians: Volunteer water quality monitoring program 1992–1994 report*. Annapolis Royal, Nova Scotia: Clean Annapolis River Project.

Robins, L. (2007). Nation-wide decentralized governance arrangements and capacities for integrated watershed management: Issues and insights from Canada. *Environments Journal, 35*(2), 4–47.

Robinson, G. M. (1997). Environment and community: Canada's Atlantic coastal action program. *The London Journal of Canadian Studies, 13,* 121–136. Retrieved from http//:www.canadian-studies.info/lccs/LJCS/Vol_13/Robinson.pdf

Sharpe, A., & Conrad, C. (2006). Community based ecological monitoring in Nova Scotia: Challenges and opportunities. *Environmental Monitoring and Assessment, 113,* 395–409. doi:10.1007/s10661-005-9091-7

Sharpe, A., & Sullivan, D. (2003). *Community science in Atlantic Canada: A survey of water quality monitoring programs*. Annapolis Royal, NS: Clean Annapolis River Project.

Shelton, A. (2013). *The accuracy of water quality monitoring data: A comparison between citizen scientists and professionals*. Halifax, NS: Saint Mary's University.

Sutherland, H. (2003). *Moving forward: An environmental management plan for the annapolis watershed*. Annapolis Royal, NS: Clean Annapolis River Project.

Timmer, D. K., de Loë, R. C., & Kreutzwiser, R. D. (2007). Source water protection in the Annapolis Valley, Nova Scotia: Lessons for building local capacity. *Land Use Policy, 24,* 187–198. doi:10.1016/j.landusepol.2005.05.005

Wagner, K. (2013). *Broken brooks 2012: Salmonidae outreach, accessibility and restoration*. Annapolis Royal, NS: Clean Annapolis River Project.

Whitelaw, G., Vaughan, H., Craig, B., & Atkinson, D. (2003). Establishing the Canadian community monitoring network. *Environmental Monitoring and Assessment, 88,* 409–418. doi:10.1023/A:1025545813057

Integrated watershed management in the Bow River basin, Alberta: experiences, challenges, and lessons learned

Judy Stewart and Mark Bennett

ABSTRACT

Alberta's Bow River is heavily engineered and hard-working, supplying water to almost 1.5 million people, while meeting the needs of hydropower, agriculture, tourism and irrigation industries upstream and downstream of Calgary. Working together since 1992, the Bow River Basin Council, a voluntary multi-stakeholder organization, with government representatives at the table, has developed watershed management plans as decision-support tools; provided a forum for relationship and trust building; shared information; and co-generated knowledge. Difficult challenges became opportunities for collaborative learning by doing. The processes involved in integrated watershed management were as important as the plans that emerged. Implementing plan objectives remains the greatest challenge.

Introduction

Water scarcity, coupled with rapid population and economic growth, can affect the health of aquatic ecosystems, and this is true in the Bow River watershed in Alberta, Canada. The population in the Bow watershed continues to grow at about 2.5–3.0% per year, while licensed withdrawals for the entire watershed remain relatively fixed. The Bow River's main stem and tributaries are closed to new surface water allocation licenses. Almost 1.5 million residents (mostly urban) depend on the Bow and its tributaries for domestic, commercial, industrial and recreational water needs. Three of Alberta's irrigation districts also rely on the Bow to deliver sufficient flows for farming and crop production southeast of Calgary, while hydropower, other industries and tourism need some of the Bow's limited supply. The average annual discharge of the Bow (at the mouth) is about 3.9 M dam³ (1 cubic decametre = 1M litres); however, discharge can vary by up to approximately +/– 50% of the annual mean (Bow River Basin Council, n.d.).

The social system and culture in the Bow watershed are inextricably connected to the health of the Bow. Water quantity and quality are managed, not just to meet the growing demands of society or the aquatic ecosystem, but to sustain the entire social-ecological system. The region has been studied as a complex, adaptive social-ecological system with open boundaries in a continual state of flux. Society, and the ecosystem that sustains it, are inextricably connected, such that feedback from the ecosystem affects society's evolution,

111

and vice versa (Stewart, 2016a; Tyler & Quinn, 2013). In this complex system, the Bow River Basin Council (BRBC) discovered that integrated watershed management is the best way to manage the Bow to sustain the instream flows necessary to support river health and biodiversity (Bow River Basin Council, 2012).

The term 'integrated watershed management' (IWM) is not commonly used in Alberta, and different definitions are found throughout the country. For example, in the Capital Regional District, British Columbia, it is defined as "an approach to watershed stewardship that aims to collaboratively manage the landscape and its development to maintain watershed function and create sustainable communities" (Capital Regional District, n.d.). In Ontario, Conservation Ontario (2013) defines it as a "process of managing human activities and natural resources on a watershed basis. This approach allows us to protect important water resources, while at the same time addressing critical issues such as the current and future impacts of rapid growth and climate change." The Government of Canada (2016) explains the IWM concept as "a multidisciplinary and iterative process that seeks to optimize the contribution of aquatic resources to the social, environmental, and economic welfare of Canadians, while maintaining the integrity of aquatic ecosystems, both now and into the future".

These definitions illustrate that IWM is a complex concept without a common definition across Canada. However, in southern Alberta there is a common understanding that watersheds are complex, dynamic social-ecological systems, where society and land, water, air and other natural resources are inextricably connected, and therefore responsive to feedback in the system (Stewart, 2016a; Tyler & Quinn, 2013).

As practiced by BRBC, IWM is focused on sustainability and is coordinated across sectors and across levels of government. Sustainability in the BRBC context is consistent with the most frequently quoted definition found in the *Brundtland Report* (World Commission on Environment and Development, 1987, p. 43):

> Sustainable development is development that meets the needs of the present without compromising the ability of future generations to meet their own needs. It contains within it two key concepts: the concept of needs, in particular the essential needs of the world's poor, to which overriding priority should be given; and the idea of limitations imposed by the state of technology and social organization on the environment's ability to meet present and future needs.

In keeping with sustainability, BRBC has multiple objectives for managing at the watershed scale that incorporate local land-use decision-making processes. Significantly, IWM is a multidisciplinary, iterative process, where each discipline provides information and methodologies to the BRBC to co-create new knowledge necessary for all stakeholders to adaptively manage water and land use. In this context, the Bow Basin Watershed Management Plan (Bow River Basin Council, 2012), referred to subsequently as the Plan, is considered an IWM plan.

For many years, the BRBC has envisioned a future where "the Bow River watershed will be conserved and protected as a fragile and unique resource and recognized as our lifeline. Multiple uses will be balanced, ensuring the needs of all stakeholders are met while recognizing that a healthy ecosystem is paramount" (Bow River Basin Council, 2015, p. 5). Since 1992, BRBC has provided an IWM forum through which stakeholders voluntarily collaborate, addressing identified shared watershed management issues in which water use and land use are connected. The wealth of knowledge co-created through BRBC is also provided to the government of Alberta as advice and practical templates (Stewart, 2014).

In Alberta, while unstated, an IWM process is required for sustainability, and this expectation is framed in *Water for Life: Alberta's Strategy for Sustainability* (Government of Alberta,

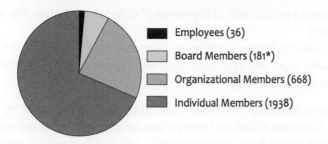

Figure 1. WPAC participation summary 2011-2012.

2003). Water for Life has three interconnected goals, in recognition that Albertans need water for healthy and prosperous lives, the sustainability of their social and cultural communities, and healthy aquatic ecosystems:

• safe, secure drinking water supply
• healthy aquatic ecosystems
• reliable, high-quality water supplies for a sustainable economy.

Water for Life established a plan with short, intermediate and long-term goals, which were reviewed and restated in 2008. These goals are to be achieved through three key strategies: knowledge and research; partnerships; and water conservation (Government of Alberta, 2003). The document recognizes that without a provincial strategy for watershed management, Albertans' ability to sustain healthy, prosperous lives would be compromised. It also acknowledges the competing interests for the same, sometimes scarce water resources, and that water management must be coordinated at the regional or watershed scale in partnership with multi-stakeholder organizations.

BRBC partners with the provincial government under Water for Life, functioning as one of 11 Watershed Planning and Advisory Councils in Alberta. More than 2500 stakeholders are directly engaged in the council network throughout Alberta (Figure 1).

IWM in the Bow watershed is challenging, especially when variability in the hydrological system leads to inevitable, but unpredictable, extreme environmental events, such as the drought of 1999–2001 and the 2013 flood, which affected the Bow and most river systems in southern Alberta. The flood demonstrated to stakeholders in the Bow watershed how land-use encroachment on the floodplain had affected aquatic health and thus human well-being. These same stakeholders continue to work together to minimize further damage to infrastructure in flood-risk areas, assess how much risk is acceptable, and determine who should bear the burden of that risk (Stewart, 2014).

Over 30 years ago, Mitchell and Gardner (1983, p. 1–4) identified several generalizations associated with 'river basin' or watershed management in Canada at that time:

• There did not appear to be any single correct or proper way to pursue river basin management.
• There was an urgent need to reduce the time to complete and implement plans.
• There was a need to broaden the focus from one of 'water' to one that included related land-based issues.

- While there was a recognition of the merits of public participation, the results had been disappointing.
- There was a need to improve communication between those involved in writing plans and those who must decide if, when and what specific recommended projects are to be implemented and funded.

Since 1983, and based on extensive case-study research on IWM by Shrubsole and Mitchell (1997), Shrubsole (2004) and many other scholars in Canada and elsewhere around the world, Mitchell (2005) further defined IWM and identified several other challenges to achieving the goals of IWM. Mitchell examined "the implications of different interpretations of a systems, ecosystem, or holistic approach related to IWM" (p. 1344–1345) in order to design institutional arrangements to facilitate IWM implementation through land-use planning. He stated that "after considerable time and effort have been allocated to [IWM], there often is relatively little action. The principal reason is that frequently the [IWM] plan has no obvious 'home' or legal basis, and therefore has low legitimacy"; and "the implication is that connecting to statutory-based land-use planning has the potential to improve the effectiveness of [IWM]". In 2006, Mitchell (2006) reiterated similar challenges of IWM, for example the typical problem of regulatory silos for water-use management and land-use management, and he stressed the need for partnerships in implementing IWM plans.

While the challenges raised by Mitchell and Gardner (1983) and Mitchell (2005, 2006) persist in the Bow watershed in 2016, to varying degrees, this paper provides some insights into the experiences of IWM practitioners in this watershed, including successes, challenges, and lessons learned. After presenting background information on the Bow watershed and BRBC, the authors illustrate where significant progress has been made towards addressing the challenges noted by Mitchell and Gardner, as well as the problems with implementing the Plan.

Background: the Bow basin and BRBC

The Bow basin is a complex, adaptive social-ecological system, where society and ecology are linked (Tyler & Quinn, 2013). The Bow Glacier, in Banff National Park, is the source of a high-elevation, fast-flowing, ice-cold river. As the Bow flows through the foothills and urban centres in the upper watershed, it is fed by several tributaries. Eventually, the Bow merges with the Oldman River to form the South Saskatchewan River, which flows into Saskatchewan. The South and North Saskatchewan Rivers subsequently merge to form the Saskatchewan River, flowing through Manitoba into the Hudson Bay. Figure 2 shows the Bow basin in southern Alberta. The Bow basin covers approximately 25,000 km², about 4% of Alberta's total area. The watershed yields about 3% of Alberta's surface water, yet provides water to over 33% of Alberta's population (Bow River Basin Council, 2012).

In the 1980s, some residents and water users downstream of Calgary were concerned about deteriorating water quality in the Bow. Responding to these concerns, the minister of environment, Ralph Klein (later to become premier), formed a task force of academics, researchers, water users, etc., to investigate the conditions and report.

The *Bow River Basin Task Force Report* (Alberta Environment, 1991) presented 33 recommendations. The first was that a multi-stakeholder council be created and report directly and regularly to the minister of environment. Thus, the Bow River Water Quality Council was

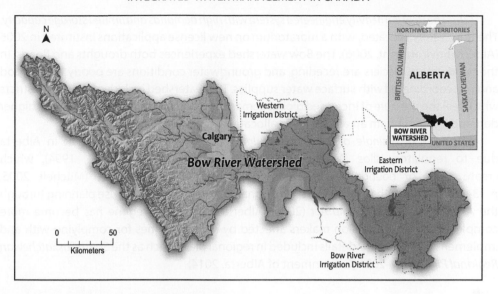

Figure 2. The Bow River Basin, Alberta, Canada.

created (by order of the Ministry of Environment), and members were appointed. The council was to promote awareness, improvement and protection of the Bow's water quality.

The council carried on until it published its first State of the River report (Bow River Water Quality Council, 1994). At that time, with emerging concerns about water over-allocation, the council's mandate was expanded to include water quantity. The reformatted organization became the Bow River Basin Water Council. In 2000, that organization formally amalgamated with a separate independent charity, the Bow River Water Quality Foundation. The new entity, the BRBC, was formed as a society and became 'arms-length' from government. Members were no longer appointed by the minister, and BRBC membership was opened to all interested parties. At the time of the merger, the organization had 49 members representing different stakeholders and organizations. By 2016, membership in the BRBC had grown to 275. Decision making is by consensus.

Under Water for Life, the BRBC was officially recognized by the provincial government as the watershed planning advisory council for the Bow watershed. As an established organization, the BRBC provided the template for the development of other watershed planning advisory councils, which have recognized roles to educate, facilitate, support and build capacity for water literacy and IWM throughout Alberta.

Summary of Bow watershed management challenges

Three overarching watershed management challenges exist in the Bow watershed.

(1)   Cumulative impacts of population and economic growth. As a result of growth in the Bow watershed, water quality has deteriorated in some river reaches (especially from increased nutrients and pesticides). Occasionally, the downstream reaches suffer from low dissolved oxygen. The watershed has experienced loss of wetlands and degraded riparian lands in high-growth urban areas to make way for residential and commercial developments (e.g. in Calgary, Cochrane and Chestermere).

(2)   The Bow is a heavily engineered system with high demands and limited storage capacity. The Bow is fully allocated, with a moratorium on new license applications instituted in 2006 (Alberta Environment, 2006). The Bow watershed experiences both droughts and floods. In the headwaters, glaciers are receding, and groundwater conditions are poorly understood and not coordinated with surface water supplies. The watershed includes irrigation districts which feel the pressure of increased global demand for food production, which means higher demand for water from an already over-allocated watershed.

(3)   Lack of a single overarching regulatory framework. IWM is difficult in Alberta due to regulatory silos (Mitchell, 2005, 2006; Mitchell & Shrubsole, 1994), which create "edge or boundary problems for institutional arrangements" (Mitchell, 2005, p. 1340–1341). With the introduction of regional-scale watershed land-use planning through the *Alberta Land Stewardship Act* (2009), Alberta's regulatory regime has become more complex, with more decision makers affected by responsibilities for complying with and implementing regulatory details included in regional plans, such as the *South Saskatchewan Regional Plan, 2014–2024* (Government of Alberta, 2014).

Challenges to practising IWM

In Alberta's practitioner community, there is not a broadly accepted set of standards and practices for IWM. Each watershed is different, geographically, geomorphologically, organizationally, politically and philosophically. The provincial government hoped to address this with publication of the *Guide to Watershed Management Planning in Alberta* (Government of Alberta, 2015) and its promotion of basic elements such as:

- Understanding current conditions in the watershed (e.g. collecting baseline data and preparing a State of the Watershed report)
- Developing a Watershed Management Plan (which includes identifying priorities and the scope and scale of planning activities; preparing and confirming support for the terms of reference; developing a communications and engagement strategy; identifying objectives, outcomes and indicators; developing, evaluating and selecting preferred management actions; and drafting and confirming support for the Plan.
- Implementing the Watershed Management Plan through building the foundation for successful implementation; establishing an Implementation Committee; and then implementing the Plan.
- Monitoring progress, evaluating, and reporting on success by tracking implementation and outcomes and then reporting on them.
- Adapting the Plan when new information becomes available.

All the above are believed to be best achieved by collaborating with stakeholders, because building trusting relationships is the key component of adaptive watershed management planning and plan implementation (Stewart, 2016a).

When the BRBC emerged from a recognized need to improve water quality downstream of Calgary, it was agreed that State of the Bow Basin reporting was an essential precursor to planning and management actions because the collection of baseline data could be assessed and monitored over time. State of the Bow Basin reporting is also essential in determining planning priorities, and has been conducted about every five years.

Successes addressing IWM challenges

One of the single greatest successes in IWM for the BRBC occurred as a result of member participation in the development of the *Approved Water Management Plan for the South Saskatchewan River Basin (Alberta)* (Alberta Environment, 2006). The cabinet approved the plan in accordance with the *Water Act* (2000). After a period of intense study and collaborative, multi-stakeholder deliberation, all those involved agreed that continued surface water allocation could no longer be supported ecologically in the Bow, Oldman, and South Saskatchewan Rivers. In response, the provincial government acted decisively, and in 2006 those watersheds were closed to new applications for surface water allocations.

Organizational capacity in IWM is variable and dependent on the staff/ volunteer ratio. In the BRBC this ratio is about 1:50, which seems to work well at this level. Revenue is a greater determinant of staffing levels than workload or volunteer numbers. Establishing sustainable revenue streams is crucial to success.

Certainly the major flooding of 2013 challenged conventional wisdom regarding IWM processes. The flood suspended or even set back some management processes and volunteer engagement opportunities. The BRBC learned valuable lessons as a result of the provincial government's response to flood mitigation, which originally proposed intervention primarily through hard infrastructure, such as dry dams. While both structural and non-structural responses are needed, the 2013 flood led to structural responses being prioritized over established IRM processes.

Lessons learned from addressing challenges to IWM

Watershed-scale land use and water management processes have proven to be fragile. The 2013 flood experience in southern Alberta led to crisis management practices, which quickly supplanted IWM processes in determining management priorities. Emergency response resulted in a rush to build engineered flood-mitigation infrastructure, setting back established best management practices, and strategies co-created through IWM processes. As of 2016, the provincial government is still actively considering major engineered solutions to mitigate Bow flooding. However, managing human development in flood-risk areas is not receiving as much attention.

While Phase 1 of BRBC's watershed management plans (Bow River Basin Council, 2008) focused on water quantity issues, the next phase focused on broader IWM concerns (Bow River Basin Council, 2012). In that context, the IWM planning process has been as important as the plan itself. During planning activities, stakeholders shared knowledge and community values, and identified problems and solutions, all resulting from discussions among people as committed to the watershed as they were to each other. Stakeholder collaboration is an essential component of IWM, and disagreement among stakeholders can create learning opportunities, e.g. environmental group members challenging industrial operations, such as forestry, oil and gas and irrigation. Working together is an exercise in voluntary self-determination, and the BRBC has chosen this route. Collaboration is not just a process; it is a crucial attitude in watershed management that participants strive to achieve.

The BRBC approach to IWM planning

Shortly after completion of the *Approved Water Management Plan for the South Saskatchewan River Basin (Alberta)* (Alberta Environment, 2006), the BRBC embarked upon Phase 1 of the Plan (Bow River Basin Council, 2008). Three key determinations were made at the outset: how to plan; what to plan; and the nature of the plan. While it is recognized that no perfect answers exist for these matters, setting benchmarks and staying with them proved successful.

The Plan Steering Committee consisted of 15–20 members, varying over the project. The committee was cross-sectoral, and for the most part the members were self-selected. If it was determined that a key sector was under-represented, a targeted recruiting effort ensued. The committee identified at least two ways to prepare a plan. One model was to identify all the integrated characteristics and then to plan for them all simultaneously – imagine this as expanding concentric ripples on a pond. A second model entailed prioritizing the same elements and aiming for significant advances on selected elements. As new elements and/ or priorities were added to the Plan, those addressed previously were updated – imagine this process as iterative, like the formation of a nautilus shell (Stewart, 2016a). The BRBC chose the latter. Thus, the Plan is an iterative creation, implemented in phases, and based on the priority determined for watershed management plan elements.

Successes and challenges of the BRBC approach

BRBC used a focus group of 23 experts to create a defensible, systematic process for determining the Plan's priorities, referred to as the Strategic Watershed Assessment Team. It identified 14 planning aspects or elements (e.g. water quality objectives, riparian health, wetland preservation). In order to establish an order of priority, each element was assessed in three ways.

The first way was called a decision support matrix. Each planning element was assessed using predetermined weighted questions administered in a web-based survey to an expert cohort of BRBC members. This was similar to a Delphi process.

Second, the same elements were reviewed in a risk analysis. This assessed the likelihood and consequences of certain aspects and/or events for each of the Plan elements. The results were then ranked from least to most risky. The third and final method involved assembling all of the known and available data (maps) on watershed sensitivity.

Some doubters suggested that, when it comes to IWM planning, a non-regulatory process would not work because the BRBC is a voluntary organization with no legal mandate and therefore cannot enforce plan implementation. As indicated by Mitchell and Gardner (1983), improved communications are necessary between those writing the plans and the land-use or water-use decision makers who implement the plans and decide whether sufficient funding is available for specific projects or programs.

It was also suggested that there was no real commitment from stakeholders or decision makers in the process. Ultimately, it was felt that these shortcomings would make success very hard to measure (i.e. selection of appropriate performance measures and measuring units). Improving performance measurement is an ongoing process. The provincial government had an opportunity to recognize the value of the BRBC's Plan, and make it a decision-support tool in the region through the *South Saskatchewan Regional Plan, 2014–2024* (Government of Alberta, 2014). Unfortunately, this did not happen, even though provincial representatives were involved in the creation of both plans.

Lessons learned for BRBC's approach

Stakeholder participation in IWM planning is crucial for the Plan's ultimate implementation. Some outcomes, the strategies to achieve them, and codes of practice in the Plan have been adopted by member municipalities in the watershed. The Calgary Regional Partnership reflected many of the IWM outcomes in its *Calgary Metropolitan Plan* (Calgary Regional Partnership, 2014), created for regional-scale land-use planning.

Perhaps the greatest lesson learned was that IWM requires management of human activities in both water and land-use planning. One cannot manage water or land independently, because they are inextricably connected (Tyler & Quinn, 2013).

In preparing the Plan, it was imperative that the BRBC design and apply a process that could build political, stakeholder and owner support as it progressed. Due to the rapid changes that may occur within the watershed (e.g. growth, industrial development, climate variation), the Plan has to be adaptive. One of the best ways to accomplish this is through identification and adoption of best management practices. Also, when planning, there needs to be integrity of process (i.e. an effort to produce a broadly accepted document). Process integrity can be fostered in the Plan because it is objectively science-based and economically feasible, and the processes of its creation were open, inclusive and transparent. If possible, peer review should be sought before a plan is launched; for example, the BRBC had the first round (water quality objectives) of the Plan reviewed by senior scientists at Environment Canada.

An inclusive and collaborative planning process engages numerous interests. It is possible that not every interest will be represented at the outset. To obtain and retain participants, two other key ingredients were essential for the Plan. First, the Steering Committee designed a process that welcomed latecomers and they could quickly be brought up to speed. Second, the process was fully web-enabled, increasing accessibility for all interested parties.

Another key lesson was the emergence of the '80% doctrine' for implementation. While about 10% of stakeholders became early adopters of the Plan, about 10% might be considered laggards. Through this doctrine (Figure 3), 80% of stakeholders were essentially implementing Plan objectives, almost saying, "We're prepared to do something, just help us with some advice on what to do." Since the project committees are populated with stakeholders, the planning process became a manifestation of stakeholders helping stakeholders (themselves) in information and advice sharing.

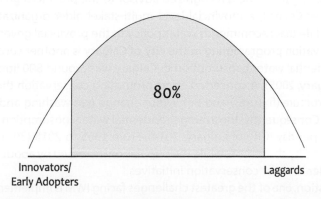

Figure 3. The 80% doctrine for plan implementation.

The Plan advises the 80% or so of its stakeholders (grouped around the mean) because it is a non-regulatory, non-binding decision-support tool. In IWM, innovators and early adopters emerge. These are stakeholders who undertake progressive management from a leadership position. One example would be the town of Banff, which embarked on a major wastewater treatment plant upgrade in the early 2000s. It was designed for performance (e.g. a 93% reduction in dissolved phosphorus) well above existing regulatory requirements. It was completed prior to finalization of the first phase of the Plan, regarding water quality objectives, promoting such improvements. Therefore, by the time the advice was offered, Banff had already met the objective.

In any plan implementation system there may be laggards who do not embrace a voluntary decision-support tool. Ultimately, there is an enduring need for a regulatory instrument to force movement by laggards, not only to meet environmental objectives but also to honour the voluntary efforts of the 80% and their beyond-compliance improvements.

BRBC's plan implementation

Implementing an IWM plan is complicated because of issues regarding organizational legitimacy and legislative authority. The Plan was created by stakeholders who self-selected to be at the BRBC planning table and voluntarily committed to the planning processes and outcomes. But the BRBC has no legislated mandate to create IWM plans that all decision makers in the watershed must subsequently adopt. As a result, some municipalities are reluctant to fully adopt the Plan, which was created by volunteers who were not elected and have no representative capacity, especially if activities require expenditure of tax dollars.

Successes and challenges of BRBC's plan implementation

One enduring virtue of the voluntary multi-stakeholder approach is that, ideally, the same participants who steer the process of developing a plan are also engaged in adopting and championing its implementation. Co-creation of a plan seems to promote ownership, and plan creators are more likely to be early adopters.

BRBC convenes quarterly forums to share information, co-generate knowledge and connect stakeholders in order to foster development of trusting relationships. Through its standing committees, BRBC is engaged in development of policy documents and submissions to government, making the BRBC a recognized advisor to the provincial government and to the Alberta Water Council, a provincial-level multi-stakeholder organization established under Water for Life that recommends water policy to the provincial government.

Water conservation programming in the city of Calgary is another success story. In the mid-1980s, residential water consumption in Calgary was around 800 litres per capita per day (City of Calgary, 2005). A concerted effort promoting conservation though regulatory (by-law), infrastructure (fixtures) and behaviour change (car washing and lawn watering) led to a decline. Consequently, the average residential water consumption in 2013 was 231 litres per capita per day (City of Calgary, 2014). From 1985 to 2015 (30 years), water consumption dropped by about 71%, while the population increased by about 390,000, or 60%. Calgary is a leader in water conservation initiatives.

Without question, one of the greatest challenges facing IWM is implementation. Rational, thoughtful and well-intentioned plans can wither from post-production neglect and inaction.

BRBC currently faces the persistent issues of the urgent need to reduce the time to complete and implement IWM plans, and the need for public engagement (Mitchell & Gardner, 1983) in all aspects of BRBC's work. BRBC's role is always a work in progress, and the board has to sustain the collaborative attitudes of stakeholders while promoting ambitious voluntary management objectives.

Lessons learned from BRBC's Plan implementation

In reviewing Plan implementation, certain observations have become apparent. Implementation is built on a foundation of sustainability, with a strong desire among stakeholders for continuous improvement, and adaptation to feedback. These observations are exemplified by shared responsibility, leading to shared stewardship. The collaborative efforts of stakeholders must be comprehensive (inclusive) for a truly integrated result to be achieved. Uncertainty should be addressed through the application of a precautionary approach. Effective implementation usually reflects effective pragmatic decision making.

Understanding the water–land nexus

Land uses and human activities in a watershed affect water quality, quantity or both. This is recognized in the mission of the BRBC (Bow River Basin Council, 2007):

> To engage all key stakeholders in protecting and improving the waters of the Bow River Basin and focusing attention on all social, cultural, economic and environmental aspects of the: riparian zones; aquatic ecosystems; quality and quantity of groundwater; and surface water effects of human activity and land use on water resources.

Successes and challenges of understanding the water–land nexus

BRBC's experience shows that appreciation of the water–land connection has evolved and matured. The focus of the original council in 1991 was instream water quality on only the Bow's main stem. This was further reflected in the 1994 State of the River report (Bow River Water Quality Council, 1994). By 2005, the State of the Basin report tacitly recognized the critically important land–water connection, literally the 'watershed'.

The second round of the Plan (Bow River Basin Council, 2012) recognized the need to broaden the focus from water to include related land-use issues. From 2005 to 2010, a deliberate transition took place in BRBC. State of the Bow Basin reports went from almost exclusively paper-based products to web-based State of the Watershed reports, accessible to all. By the time the 2010 State of the Watershed Summary and associated project were completed, 5 of 21 key indicators were landscape measures, to complement the 16 indicators focused on water quality, water quantity and aquatic ecosystem health.

As the notion of the water–land connection was better understood, it quickly became apparent to BRBC stakeholders that the necessary data-sets for truly integrated analysis, modelling and management were not readily available in Alberta. While recognition of the connection spread quickly, data collection did not follow, although the provincial government is making headway regarding groundwater mapping, wetland policy development, and recognition that water and land management are inextricably connected.

Municipalities have responsibility for land-use management of private lands in the Bow basin, while the province regulates and controls the diversion and use of water and approves activities affecting water quality. The provincial government also manages land use on public lands, while the federal government is involved in management plans for Indian reserves and national parks. As the Bow basin includes all such land types, legal pluralism exists (Stewart, 2016b), often making it difficult to determine which decision makers are responsible for implementing co-created plans. While this is definitely a challenge, BRBC is fortunate to have membership from all levels of government to work through the complicated policy, legislation and regulations.

Following Mitchell and Gardner's advice from over 30 years ago, BRBC stakeholders have established an implementation committee, who are actively engaged in communicating between those involved in writing the Plan and those who must decide if, when and what specific recommended projects are to be implemented and funded.

Lessons learned from understanding the water–land nexus

Broad understanding of the complexity of the water–land nexus has been confounded by the 2013 flood event, which has been referred to as the most expensive disaster in Canadian history (it was in the CAD 6–8 billion range). However, not everyone (neither the public nor decision makers) seems to appreciate that while it was water that caused the flood (the hydrological event), it was land-use practices, past and present, that caused the damage – i.e. the socio-economic impacts (Stewart, 2014). For example, homes and industry in the floodplain were flooded and damaged. The same homes and industry, built out of harm's way, would not have been affected.

Public participation

Reflecting on Mitchell and Gardner's ideas, while there is growing public interest in watershed management in the Bow watershed, the results of public engagement have sometimes been disappointing. BRBC is now a recognized partner with government in co-creating new knowledge for watershed management, but the general public in the watershed may have little awareness of the partnership or the system dynamic.

Successes and challenges in public participation

In 2014, two major public-engagement events provided legitimacy and encouragement to BRBC members. First, the provincial flood mitigation task force worked closely with BRBC, recognizing it as a voice for the Bow and its tributaries. Second, following closely on the heels of flood mitigation, the Alberta Water Smart (2014) Room for the River project called on BRBC as an organization, and its stakeholder members, to design an appropriate system for the Bow and Elbow Rivers, the latter a major tributary emptying into the Bow at Calgary. These two activities were major successes, legitimizing the BRBC in planning and advising decision makers, at least regarding flood mitigation.

Understanding the importance of public engagement and actually developing appropriate mechanisms to engage with the public are two different things. The watershed-management social network is highly centralized and one for which information becomes highly refined, but often it does not reach the general public (Stewart, 2016a). The flood in 2013 certainly provided impetus for people to learn more about flood mitigation, but the opportunity provided by that crisis to engage with the general public has largely passed. Membership in the BRBC remains steady, with an influx of new members as old members leave. However, a strong core of members actively engage in IWM planning and fully embrace the important functions of BRBC as a multi-stakeholder organization. BRBC performs a critical function in the co-creation of values for regional-scale watershed management and policy development (Stewart, 2016a, 2016b).

Lessons learned from public engagement

To promote public participation, processes must be transparent, with clear accountability. In addition, they must be cost-effective and fair. The public must see various planning initiatives (at the local, regional and provincial scales) that are coordinated, integrated and inclusive. Ultimately, the public needs assurance that co-created IWM plans will be used by land-use and water-use decision makers throughout the Bow basin.

Conclusions

The Bow River watershed is a complex social-ecological system, with an active and stable multi-stakeholder organization actively pursuing IWM. The challenges to IWM identified by Mitchell and Gardner (1983) and by Mitchell (2005, 2006) persist, but in the Bow basin these challenges have led to success stories and celebrations.

There are many noteworthy successes (e.g. learning, submissions, fundraisers, celebrations, workshops and road shows), but perhaps the most significant is the sustainability of the BRBC as a recognized voice for the river in regional and provincial IWM. Across BRBC's several incarnations as an organization, a core group of volunteers has led the complex process of watershed planning. They have developed a valuable network of stakeholders who trust one another and who readily and voluntarily come together to tackle tough watershed management issues, as for example the BRBC did to create a made-in-Alberta Room for the River process (Alberta Water Smart, 2014) .

Two key lessons learned by the BRBC are that the process of planning is as important as the product that emerges, and that implementation remains the greatest challenge to IWM. However, when a group of intelligent, highly motivated stakeholders collaborate in IWM, agreement usually is reached on a primary desired outcome – "to have the best managed watershed in the world" (Bow River Basin Council, 2015). While there may not be one pre-scribed right way to roll out IWM, it is certainly possible to collaborate, build trust and use system feedback to adapt and evolve over time.

Sometimes the distance between where you start in IWM and where you want to go can seem considerable and formidable. After years of concerted effort, it can seem that little progress has been achieved. It is imperative that participants in the process be regularly reminded of what has been done, and not just asked to focus on what is left to do. BRBC is committed to always reporting on good news, and regularly celebrating success. Members realize that if they were not involved, much less would have been accomplished since 1992 in managing the Bow watershed.

In IWM, it is impossible to fail; it is only possible to limit success.

Disclosure statement

No potential conflict of interest was reported by the authors.

References

Alberta Environment. (1991). *Bow river basin task force report*. Unpublished.

Alberta Environment. (2006). *Approved water management plan for the South Saskatchewan river basin (Alberta)*. Edmonton, AB: Alberta Environment. Retrieved from http://esrd.alberta.ca/water/programs-and-services/river-management-frameworks/south-saskatchewan-river-basin-approved-water-management-plan/documents/ApprovedWaterManagementPlanForThe-SSRB.pdf

Alberta Land Stewardship Act, SA. (2009). c.A-26.8. Retrieved from http://www.qp.alberta.ca/1266.cfm?page=A26P8.cfm&leg_type=Acts&isbncln=9780779778171

Alberta Water Smart. (2014). *Room for the river pilot in the Bow river basin advice to the government of Alberta*. Retrieved from file:///C:/Users/Judy/Downloads/Room%20for%20the%20River%20Pilot%20in%20the%20Bow%20Basin_Report_2014-12-19.pdf

Bow River Basin Council. (2007, January 12). *A.2 BRBC vision (policy adopted by the BRBC Board)*. Unpublished.

Bow River Basin Council. (2008). *Bow basin watershed management plan: Phase 1: Water quality*. Calgary, AB: Bow River Basin Council. Retrieved from http://wsow.brbc.ab.ca/reports/BBWMP.pdf

Bow River Basin Council. (2012). *Bow basin watershed management plan 2012: land use, headwaters, wetlands, riparian lands, water quality*. Calgary: Bow River Basin Council. Calgary, AB: Bow River Basin Council. Retrieved from http://www.brbc.ab.ca/index.php/resources/bbwmp

Bow River Basin Council. (2015). *Bow river basin council business plan April 1 2015 – March 31 2018*. Retrieved from http://www.brbc.ab.ca/brbc-documents/board-documents/244-brbc-business-plan-2015-2018

Bow River Basin Council. (n.d.). Retrieved from: http://www.brbc.ab.ca/

Bow River Water Quality Council. (1994). *Preserving Our Lifeline: A Report on the State of the Bow River*. Unpublished. Retrieved from http://brbc.ab.ca/resources/publications

Calgary Regional Partnership. (2014). *Calgary metropolitan plan*. Calgary, AB: Calgary Regional Partnership. Retrieved from http://calgaryregion.ca/cmp/bin1/pdf/CMP.pdf

Capital Regional District. (n.d). *Watershed stewardship and integrated watershed management page*. Retrieved from https://www.crd.bc.ca/education/our-environment/watersheds/integrated-watershed-management

City of Calgary. (2005). *Water efficiency plan – 30 – in – 30, by 2033*. Retrieved from http://www.calgary.ca/_layouts/cocis/DirectDownload.aspx?target=http%3a%2f%2fwww.calgary.ca%2fUEP%2fWater%2fDocuments%2fWater-Documents%2fwater_efficiency_plan.pdf&noredirect=1&sf=1

City of Calgary. (2014). *2013 Water report: Counting on our water from the river to the tap and back*. Retrieved from http://www.calgary.ca/_layouts/cocis/DirectDownload.aspx?target=http%3a%2f%2fwww.calgary.ca%2fUEP%2fWater%2fDocuments%2fWater-Documents%2fWater-Report.pdf&noredirect=1&sf=1

Conservation Ontario. (2013). *Integrated watershed management*. Retrieved from http://conservationontario.ca/what-we-do/what-is-watershed-management/integrated-watershed-management

Government of Alberta. (2003). *Water For Life: Alberta's Strategy for Sustainability*: Edmonton, AB: Government of Alberta. Retrieved from http://www.waterforlife.gov.ab.ca/docs/strategyNov03.pdf

Government of Alberta. (2014). *South Saskatchewan Regional Plan, 2014-2024*. Edmonton, AB: Government of Alberta. Retrieved from https://landuse.alberta.ca/LandUse%20Documents/South%20Saskatchewan%20Regional%20Plan_2014-07.pdf

Government of Alberta. (2015). *Guide to Watershed Management Planning in Alberta*. Edmonton, AB: Government of Alberta. Retrieved from http://www.waterforlife.alberta.ca/documents/GuideWatershedPlanningAlberta-2015.pdf

Government of Canada. (2016). *Environment Canada: Integrated Resource Management*. Retrieved from http://www.ec.gc.ca/eau-water/default.asp?lang=En&n=13D23813-1

Mitchell, B. (2005). Integrated water resource management, institutional arrangements, and land-use planning. *Environment and Planning A, 37*, 1335–1352.

Mitchell, B. (2006). IWRM in practice: lessons from Canadian experiences universities council on water resources. *Journal of Contemporary Water Research & Education, 135*, 51–55.

Mitchell, B., & Gardner, J.S. (Eds.). (1983). *River Basin Management: Canadian Experiences*. Waterloo, Ontario: University of Waterloo, Department of Geography Publication Series No. 20.

Mitchell, B., & Shrubsole, D. (1994). *Canadian water management: Visions for sustainability*. Cambridge, ON: Canadian Water Resources Association.

Shrubsole, D. (Ed.). (2004). *Canadian perspectives on integrated water resources management*. Cambridge, ON: Canadian Water Resources Association.

Shrubsole, D., & Mitchell, B. (Eds.). (1997). *Practising sustainable water management: Canadian and international experiences*. Cambridge, ON: Canadian Water Resources Association.

Stewart, J. (2014) *Summary of the workshop feedback: Flooding in rural areas perspectives, strategies, and way forward*. Unpublished proceedings. Retrieved from http://www.brbc.*ab*.ca/index.php/about-us/committees/legislation-and-policy

Stewart, J. (2016a) *A reflexive legal framework for bridging organizations in regional environmental governance and management*. Doctoral Dissertation, Faculty of Environmental Design, University of Calgary. Unpublished.

Stewart, J. (2016b). Reflexive legal processes for environmental bridging organizations in the Calgary region. Canadian institute of resources law. *Resources* 119: 1. Retrieved from http://prism.ucalgary.ca/handle/1880/51362

Tyler, M.E., & Quinn, M. (2013). Identifying social-ecological couplings for regional sustainability in a rapidly urbanizing water-limited area of western Canada. In C.A. Brebbia (Ed.), *Wessex Sustainability Development and Planning VI* (pp. 175–191). Southampton, UK: WITPress.

Water Act, R.S.A. (2000). c.W-3. Retrieved from http://www.qp.alberta.ca/1266.cfm?page=W03.cfm&leg_type=Acts&isbncln=9780779787272

World Commission on Environment and Development (1987). *Our common future*. Oxford: Oxford University Press.

The Northeast Avalon Atlantic Coastal Action Program: implementing integrated watershed management in Newfoundland and Labrador

Kailyn Burke

ABSTRACT

The Northeast Avalon Atlantic Coastal Action Program (NAACAP) is a non-profit organization dedicated to protecting watersheds and coastal environments through research, education, community engagement and strong cross-sectoral partnerships. NAACAP's mission statement establishes the principles of integrated watershed management, which drive the efforts of the organization: local initiatives; partnerships and collaboration; watershed basis; and aquatic health. Central to NAACAP's successes are strong partnerships with industry, other non-profits and local, provincial and federal governments. This is key to overcoming the ongoing anthropogenic and organizational challenges faced by NAACAP in the application of integrated watershed management in the Northeast Avalon region of Newfoundland and Labrador.

Introduction to NAACAP watershed management

Prior to the 1990s in Newfoundland and Labrador, water governance was a responsibility shared between the provincial and federal governments. The provincial government managed inland water quality, flood control and wildlife and fish habitat, whereas the marine and coastal environments were the responsibility of the federal government. In 1991, following the development of the federal government's Green Plan (Government of Canada, 1990), an urgent need was recognized to empower local communities to restore coastal environments in order to sustain coastal communities (Gardner Pinfold Consulting Economists., 2005).

The Northeast Avalon Atlantic Coastal Action Program (NAACAP) was founded in 1993 (it was first called the St. John's Harbor Atlantic Coastal Action Program) with funding through Environment Canada's Atlantic Coastal Action Program (ACAP) as part of the Green Plan. A total of 13 organizations originally participated in the programme throughout Atlantic Canada, including 2 in Newfoundland. ACAP was intended to be a means of empowering local communities in identifying and addressing environmental and development challenges. ACAP also marked a shift in water governance in Atlantic Canada towards a more integrated watershed management (IWM) approach (Gardner Pinfold Consulting Economists.,

2005). In particular, this approach demonstrated a fundamental shift towards a more 'bottom-up' and collaborative decision-making process that involved diverse stakeholders in the Atlantic region (Environment Canada, 2003). NAACAP was formed under ACAP and was encouraged to collaborate with industry, all levels of government, and other stakeholder organizations. As the impacts of development extended across municipal boundaries, there was seen to be a need for greater collaboration among municipalities, senior levels of government and the private and non-profit sectors to effectively address the issue of long-term conservation in the Atlantic region.

Currently, and as described in the next section, NAACAP operates as a non-profit organization. "Northeast Avalon ACAP is a citizens' organization that works with all sectors of the community, including all three levels of government, to protect and enhance the aquatic environmental quality of the watersheds and coastline within the ACAP project area" (Northeast Avalon Atlantic Canada Action Program [NAACAP], 2013, 2). This mission statement establishes the principles of IWM, which drive the efforts of the organization: local initiatives, involvement and education; partnerships and collaboration; watershed unit; and aquatic health. In its work, NAACAP applies an IWM approach, focusing on the watershed as a whole and the elements that affect it both internally and externally. NAACAP defines IWM as the process of implementing conservation efforts centred round the recognition and treatment of anthropogenic changes to watershed ecosystems as a whole. NAACAP's scope is necessarily broad and includes diverse projects, which address both large- and small-scale elements, such as sanitation and waste management; soil erosion and sediment control; the planting of trees, shrubs and herbaceous materials; invertebrates and fish; and community engagement activities. While NAACAP does not focus specifically on land animals and birds in the region, it regularly works with other non-profit organizations that do, including the Newfoundland and Labrador chapters of Ducks Unlimited Canada, and the Canadian Parks and Wilderness Society. As an organization, NAACAP believes that the protection of the unique and numerous watersheds across Northeast Avalon is important to the local quality of life. Watersheds provide areas of complex ecosystems rich with aquatic life, animals and plants; they act as natural areas for recreational activities such as hiking and fishing; and they act as natural lungs and filters for their surroundings.

Since its inception, involvement and input from the community has been a priority in all of NAACAP's projects and initiatives. Examples of community involvement include volunteerism, workshop participation, and efforts towards incorporating citizens' scientific research. It is through partnerships and community engagement that watershed conservation efforts by NAACAP have been successful and continue to thrive in the region. The aim of this case study is to assess the successes and challenges faced by NAACAP in implementing its IWM programme.

Northeast Avalon region

The Northeast Avalon region covers nearly 1160 km^2 of the Avalon Peninsula in eastern Newfoundland and Labrador that runs north of the Witless Bay Line to Cape St. Francis (Figure 1). The region encompasses 15 municipalities, with populations ranging from 130 to 106,172 (Table 1). Northeast Avalon has over 100 rivers and streams, which combine to form 78 watersheds. Watersheds in the region face varying pressures from different land-use activities. There are approximately 200,000 people living in the region, with nearly 38% of the

Figure 1. Northeast Avalon region.

Table 1. Municipalities in Northeast Avalon (Statistics Canada, 2011)

Municipality	Population (2011)	Geographical area (km²)
Holyrood	1,995	n/a
Conception Bay South	24,848	59.27
Paradise	17,695	29.24
Portugal Cove–St. Philips	7,366	57.35
Bell Island	130	31.5
Bauline	397	n/a
Pouch Cove	1,866	n/a
Flatrock	1,457	n/a
Torbay	7,397	34.88
Logy Bay–Middle Cove–Outer Cove	2,098	n/a
St. John's	106,172	446.06
Mount Pearl	24,284	15.75
Petty Harbour–Maddox Cove	924	n/a
Bay Bulls	1,283	n/a
Witless Bay	1,179	n/a

total population of the province in its capital, the city of St. John's (Statistics Canada, 2011). The region has the highest rate of urbanization and development in the province, resulting

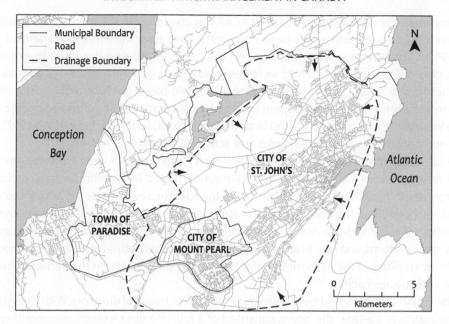

Figure 2. General service area of the St. John's sewer system (City of St. John's, 2003).

in diverse infrastructure and environmental challenges. With the significant variance in population size amongst the municipalities, the infrastructure capacity varies drastically as well. Some municipalities, Witless Bay, for example, do not have paid elected officials, but run on the leadership of volunteer councillors.

Beginnings: purpose and scope of ACAP

ACAP was established in 1991 by Environment Canada in response to greater demand for public involvement in decisions relating to the environment and increasing concerns about water quality in Atlantic Canada. ACAP was a community-driven strategy designed to build the capacity of ecosystem-based coalitions of stakeholders to prioritize and drive restoration and sustainability initiatives in their local environments (Gardner Pinfold Consulting Economists., 2005). The intention was to address four broad areas of concern identified in the Canada Water Act (Environment Canada, 1985): clean water; atmospheric depositions; toxics; and natural habitats (Environment Canada, n.d.). The programme was founded on four key watershed management principles – recognizing ecosystems as communities, knowledge generation, capacity building and direct action – which subsequently became the foundation of the organization's IWM approach (Environment Canada, 1998). Since 1991, ACAP groups have been established throughout New Brunswick, Prince Edward Island, Nova Scotia, and Newfoundland and Labrador (Environment Canada, 2003). One of the primary objectives of the programme was to develop collaborative watershed-based community plans for the management of the coastal environment in 13 project areas throughout the Atlantic Provinces, including St. John's Harbour.

St. John's Harbour ACAP was founded in 1993 by a group of citizens and government stakeholders under a federal (Atlantic Canadian) initiative developed and funded by

Environment Canada. St. John's Harbour ACAP was created with the mandate to develop a community-directed, consensus-based comprehensive environmental management plan (CEMP) to address the sewage flowing into the harbour. Utilizing the same foundational principles of IWM activities, CEMPs were designed to look at the watershed as a whole and the factors that impact it, and to encourage watershed-based decision making. The organization initially focused on developing sewage treatment for the St. John's region; the initiative was referred to as the St. John's Harbour Clean-Up (St. John's Harbour ACAP, 1997). This was very consistent with the principles and activities envisaged by ACAP.

Since the beginning, the members of St. John's Harbour ACAP were a diverse group of local residents from a broad range of backgrounds, united by a common concern for the environmental health and aesthetic features of the harbour. Sewage was the primary concern of the Harbour Clean-Up, but the initiative also focused on eliminating the other waste and debris that was being deposited in the harbour through the sewage systems, such as sanitary products, condoms, and paints and solvents. This additional debris not only diminished the aesthetic qualities of the harbour but also had significant impacts on the harbour environment and species habitat.

The organization was led by a 21-member, volunteer board of directors. With the support of Environment Canada, the board consisted of a four-member executive committee and nine other voting directors, elected annually from and by the membership. The board also included a non-voting group of advisory directors, representing each of the three levels of government (municipal, provincial and federal).

By 1993, untreated municipal sewage was the major source of pollution in St. John's Harbour, as there were relatively few heavy industries or large manufacturing activities in the watersheds contributing to the harbour. The main focus of St. John's Harbour ACAP was to build capacity to treat the 120,000 m^3 of raw sewage and stormwater entering the harbour each day (St. John's Harbour ACAP, 1997). During wet-weather storm events, four to five times the daily outfall amount from the sanitary sewage system went into the harbour. Since the wastewater was not treated or disinfected, the harbour was contaminated with potentially pathogenic bacteria that could cause ear and gastrointestinal infection from bodily contact with harbour water (City of St. John's, 2003). There was also a widespread and varied array of light industries, commercial operations, and organizations that had the potential to be significant contributors of pollution from accidental spillage into the rivers flowing through the city. The focus on sewage treatment was a direct result of the founding purpose of the organization, which was to develop an effective management plan to ensure the sustainability of the coastal community in and around St. John's Harbour.

The cumulative impacts on St. John's Harbour's contributing watersheds needed to be considered. At the time, effluent discharges from commercial and industrial sources were only subject to provincial regulations and the general prohibitions of the federal Fisheries Act (Government of Canada, 1985) if the discharges were directed to receiving waters. As a result, in the St. John's area commercial discharges were directed to the sanitary sewers to avoid penalty, but still ultimately discharged untreated into the harbour.

The project area defined in the mandate of St. John's Harbour ACAP was land which naturally or artificially drained into St. John's Harbour, Quidi Vidi Harbour, or St. John's Bay (including its constituent bays and coves). This area included the cities of St. John's and Mount Pearl, as well as part of the town of Paradise (Figure 2). The total population of the harbour sewer area in 1996 was 127,370, including households in each of the three municipalities (City of St. John's, 2003, ES-3-5).

The early activities of St. John's Harbour ACAP focused on accumulating and distributing information on the environmental quality of the harbour. This was followed by the development of an appropriate action plan and consensus building in the community to deal with the problem of untreated sewage. Consultations and workshops were held wherever possible with the public and other stakeholders. Where gaps in existing knowledge were identified, such as the unknown effects of raw sewage on aquatic life in the harbour and sediment quality to appropriately address the problem, the organization sought funding to carry out the necessary research to ensure that all decisions would be based on sound scientific evidence.

In addition to the preparation of the CEMP, St. John's Harbour ACAP gathered and distributed information pertinent to the efforts to clean up St. John's Harbour. The information campaign was intended to encourage participation and buy-in from the public to help attain the programme's objectives. The decision to include and continuously interact with public and private agencies, as well as all levels of government, was made to encourage cooperation and consensus building regarding the clean-up of the harbour and its surrounding environment. The plan for protecting and rehabilitating the harbour and its related marine, estuarine and tributary areas was to be community directed and scientifically defensible at all times.

St. John's Harbour ACAP received core funding from Environment Canada upon its inception. Some project-based support was also provided over the years from the Department of Fisheries and Oceans and the provincial government. In its time as St. John's Harbour ACAP, from 1993 to 2005, the operating budget of the organization varied greatly from year to year, ranging from CAD 45,000 to CAD 140,000. Between 1993 and 1997, Environment Canada provided core funding to develop the long-term strategies to manage the local ecosystem with the CEMP for the harbour and complete individual projects. This shifted to project-based funding in 1998 (Gardner Pinfold Consulting Economists., 2005). During this time, the organization was financial stable, as the majority of operating and project costs were provided by Environment Canada.

One of the long-term goals of St. John's Harbour ACAP was to create a remedial action plan to address the issues at hand. The components of the remedial action plan were to include monitoring of the water quality of the rivers that flowed into the harbour and the seabed sediment quality under the harbour, and continued public awareness campaigns about the watershed and harbour ecosystem through education on hazardous material disposal and other related water issues. Although a comprehensive remedial action plan was not developed, the key aspects of the plan's framework continue to influence projects to this day.

Transition years

On 4 November 2002, Prime Minister Jean Chrétien came to St. John's to announce the commitment of CAD 31 million from the federal government towards the construction of a wastewater treatment plant for the harbour (Environment Canada, 2003). This matched the CAD 31 million funding previously committed by the Newfoundland and Labrador provincial government and the CAD 31 million collectively contributed by the surrounding municipalities. The municipal contribution was made up of CAD 27.2 million from the city of St. John's,

CAD 3.2 million from the city of Mount Pearl, and CAD 748,000 from the town of Paradise (*The Telegram*, 2013b). With this announcement of tri-level government funding to bring sewage treatment to St. John's Harbour at an estimated cost of CAD 93 million, St. John's Harbour ACAP was able to shift its mandate to address other environmental problems in the urban and surrounding regions. In 2005, St. John's Harbour ACAP revised its focus and expanded its geographical area of interest to include all the watersheds of Northeast Avalon. To reflect these changes, the organization changed its name to Northeast Avalon Atlantic Coastal Action Program.

Water quality in St. John's Harbour improved after the Riverhead Sewage Treatment Plant was built in 2009. From 2009 to 2012, as testing by the city of St. John's confirmed, coliform levels dropped by 90% on average. However, ammonia levels rose by 70%, and nitrogen and phosphorous levels stayed the same, as did acidity. While the visual signs of excrement decreased, not all sewage outfalls were able to be redirected to the treatment plant immediately, due to limited capacity (*The Telegram*, 2013a). The final sewage outfall at Temperance Street, which accounted for 30–40% of St. John's sewage, was finally redirected for primary treatment in May 2015 (*The Telegram*, "Harbour bubble gone after more than a century", 2015).

Governance and structure

The structure of governance has changed very little since the organization shifted focus to include the whole Northeast Avalon region, because the formula has remained successful. The volunteer board that governs the organization presently has 13 directors, with an executive committee of four members. Directors are elected annually, at the annual general meeting, by NAACAP members and serve a term of two years. The executive committee is nominated and selected by the elected directors annually.

Representatives of all levels of government are strongly encouraged to sit on the board in a non-voting, advisory capacity. *Ex officio* government representatives are supplied on an ongoing basis as in-kind contributions from the respective departments. As of 2014, Mount Pearl was the only municipality with an *ex officio* representative on the board. Federally, Environment Canada and the Department of Fisheries and Oceans each supply an advisory representative to the board. Provincially, representatives of the Department of Environment and Conservation and the Department of Natural Resources also regularly attend meetings and provide support in an advisory capacity.

Currently, the board conducts meetings monthly, with working subcommittee groups meeting as required to address a variety of issues related to current projects, public outreach and the internal governance of the organization. Standing committees include the Finance Committee, which oversees financial matters, and the Science and Technical Committee, which provides technical expertise. These committees take a leadership role in their respective areas and include reputable experts from the region, who provide advice and direction to staff.

NAACAP currently has two staff members, an environmental technologist and an outreach and office coordinator, who manage day-to-day operations. Prior to 2011, NAACAP's staff consisted of an environmental technologist and an executive director.

NAACAP's IWM approach

Incidentally, the shift towards working on a watershed basis throughout the entire region corresponded with concerted efforts at 'thinking regionally', supported by efforts from non-governmental organizations (NGOs) such as Municipalities Newfoundland and Labrador and the former Northeast Avalon Regional Economic Development Board. Using the watershed boundaries rather than municipal boundaries is critical for better understanding and protecting the waterways in the Northeast Avalon region, as well as garnering the appropriate political response to multi-jurisdictional issues. The purpose of ACAP was to create an ecosystem-based organization at the watershed level, an approach necessary to manage the expanding and diverse anthropogenic challenges in the region.

The Northeast Avalon region houses diverse challenges, requiring customized conservation and research efforts to address distinct environmental risks and vulnerabilities. While some of the watersheds in the region flow through industrialized and residential-development areas, others flow through agricultural land. The goal of NAACAP is to monitor and conserve the environmental health of all watersheds in the project area to attain sustainability and conservation in the watersheds of the region (Northeast Avalon ACAP, 2006). Efforts include diverse remediation efforts and water quality monitoring, which offer extensive opportunities for citizen scientists to participate in conservation efforts such as clean-ups, planting initiatives and even borrowing equipment to conduct *in situ* water testing.

The strategy of NAACAP remains focused on implementing an IWM approach that includes a coordinated, community-directed, scientifically defensible approach to watershed conservation to improve the quality of all water resources in Northeast Avalon. As demonstrated by the wide-ranging projects described below, gathering and distributing science-based evidence is at the core of NAACAP's work. The purpose of this is to build community engagement and consensus, and inform all levels of government about effective and cooperative protection of the aquatic environment.

A decade of new projects (2005–2015)

An integral part of effective watershed and coastline management is community engagement and knowledge transfer. For this purpose, NAACAP offers multiple mechanisms to enable the community to take ownership of the watersheds in the region. NAACAP hosts workshops and annual general meetings, produces and distributes publications, and organizes river clean-ups and plantings to increase the public's knowledge and understanding of proper watershed and coastal management. NAACAP also encourages direct water and soil conservation throughout all sectors in the Northeast Avalon region, including the primary sector (agriculture, mining and forestry), the secondary sector (processing, manufacturing and construction), the tertiary sector (service provision) and the quaternary sector (academia and education). This section will highlight a few of the projects and workshops NAACAP has been involved in over the last decade.

NAACAP's projects are primarily funded through grants from various federal and provincial departments, with some matching funds from the commercial sector (NAACAP, 2015). As a result, the project reports are not only used to disseminate information to the public, they are also directly received and reviewed by government agencies to provide policy makers with the data needed to make well-informed policies.

The publications created by NAACAP are tailored to various audiences, including the general public, non-profit organizations, government and the private sector, to promote understanding of key issues in the region. For example, NAACAP's report *Freshwater Joys* was developed for the public audience and encourages water conservation (NAACAP, 2009). The report provides guidelines on various actions the public can take to help NAACAP achieve its goals, such as lowering water usage and upgrading plumbing to conserve water at home. It also provides advice on a variety of topics, including erosion prevention, composting, recycling and septic system maintenance, to promote community responsibility and ownership of the health and protection of their watershed (Bartellas, 2010).

Workshops inform public and private stakeholders of the important elements of IWM, looking at the effects of commercial and community activity on the watershed as a whole. They have been held on a number of topics, such as responsible erosion and sediment control, and watershed sustainability. NAACAP's workshops provide opportunity for cross-sector discussion amongst stakeholders and provide a forum for positive and productive discourse. In June 2014, NAACAP hosted and facilitated the Working Together for Responsible Erosion and Sediment Control workshop in St. John's, in collaboration with the Newfoundland and Labrador Environmental Industry Association. Seventy-four attendees from groups representing all three levels of government, industry, students and civil society were present. Six speakers from government and industry presented their perspectives and experiences with sedimentation and erosion, with networking opportunities built into the structure of the workshop. One of the main achievements of the workshop, as revealed by participant feedback, was that it clarified the theory and regulations pertaining to erosion and sedimentation. Workshop participants also identified the importance of collaborative meetings where stakeholders in the environmental industry have the opportunity to exchange current information and ideas (Favaro, 2014).

Through workshops and meetings, NAACAP facilitates a deeper discussion between industry and the public about the ecosystem as a whole and promotes long-term, sustainable planning around local development. This is achieved through increased understanding of the short- and long-term effects of development and commercial activities on the watershed as a whole. NAACAP also organizes scientific research and shares the results with the public, municipal and provincial planners, and decision makers. As part of this research, NAACAP is interested in tracking the changes in water quality over time in an increasingly urbanizing area. Since 2003, NAACAP has collected water quality samples and data on flora and fauna in selected watersheds of the Northeast Avalon region.

Since 2005, NAACAP has performed *in situ* water quality testing using a YSI sonde to monitor numerous projects across the region measuring water temperature, pH, dissolved oxygen, specific conductivity, salinity and total dissolved solids. Unfortunately, as a result of a project-based monitoring approach, the data collected up to 2014 were piecemeal and came from select watersheds as a component of broader projects, and therefore detailed comparisons of these data are not possible. However, in 2014, NAACAP began water quality monitoring across the region, testing at 67 different sites in 41 of the 78 watersheds in the region. Each site is visited four times throughout the spring, summer and fall months. This is accomplished with the support of citizen scientists.

Through standardized training and detailed protocols, community members of all backgrounds are able to assist with and ensure continual water quality monitoring. Mandatory online training prior to using the equipment through CURA H_2O's Wet-Pro certification ensures quality assurance amongst community scientists and partner organizations. Additionally, all volunteers receive *in situ* training from NAACAP staff and intermittent check-ins in the field for quality control. This is part of a three-year initiative designed to understand the current conditions of the waterways in the region and trends and significant changes within the watersheds over time.

The results of NAACAP's studies are disseminated through various media, such as reports, community presentations, scientific papers, and brochures (Northeast Avalon Atlantic Canada Action Plan (NAACAP), n.d.). NAACAP's research projects also examine many issues affecting watersheds, such as sedimentation, erosion, invasive species, land use and zoning, and human behaviour. In 2005, NAACAP initiated the Regional Watershed Survey: Nut Brook Drainage Basin, St. John's Stream Analysis of a River System in a Local Industrial Zone, which focused on a small stream called Nut Brook, located on the outskirts of St. John's 'brown zone' of industrial use. The impacts of the industrial effluent were felt downstream, which caused the public and businesses in the area to address the problem. This is especially important as Nut Brook is the headwaters of the Kelligrews River, which flows through residential areas in the town of Conception Bay South. The study made several key recommendations: (1) water quality in Nut Brook should be regularly monitored; (2) regulatory agencies should inspect the industrial activity on Incinerator Road to confirm compliance with environmental and sanitary laws; and (3) those who may be involved with and/or affected by the contamination, or those who can assist with the clean-up, monitoring, or enforcement of the activity on Incinerator Road, should be informed of the present state of water quality (Ficken, 2006). As a result, NAACAP has undertaken baseline water quality monitoring in the area as part of the regional water quality monitoring project, and through participation on the Incinerator Road Environmental Committee is in regular contact with those who may be affected by contamination and provides input on activities to minimize contamination. Regulatory agencies investigate and enforce compliance with environmental policy.

In summary, NAACAP's projects aim to increase knowledge and understanding of the watershed and coastal areas. Conservation is a constant theme in the planning of activities, project deliberations and outcome assessments regarding NAACAP projects. By following the *Freshwater Joys* guide, being involved in one of NAACAP's many river clean-ups, or participating in a workshop, individuals are informed in an effective way and learn the values of water resources and how their own practices play a direct role in their protection. By understanding what is in the watershed and the harm which is being done to it – such as in the case of Nut Brook – local people and industry can be more efficient planners and advocates in achieving watershed management goals. These opportunities enable individuals to start thinking more about conservation and curb the practices which impact water resources negatively.

Success with community partnerships

Since NAACAP's inception, partnerships and coalitions have been central to the success of its conservation efforts. Some of NAACAP's partnerships began as part of the formal structure of the organization. As a member of the former ACAP, which later evolved into the Atlantic Ecosystems Initiative (AEI), the organization was part of a community of 14 non-profit organizations in the four Atlantic Canada provinces. The organizations operated independently but were formally linked through shared goals under the ACAP umbrella (Environment Canada, 2003).

The shift to AEI marked an end to a formal relationship amongst ACAP organizations and funding, and a significant shift to project-based funding opportunities with broader funding priorities (Environment Canada, 2015). In the past, the leaders of the ACAPs held meetings and strategic brainstorming sessions. While this occurs less frequently now that the formal relationship of the ACAP/AEI umbrella no longer exists, strong partnerships still endure among the group of ACAPs. In 2015, with the multi-regional government funding opportunities in Atlantic Canada due to changes in AEI's funding structure, NAACAP and other former ACAPs have initiated various partnership projects, including the Canadian Aquatic Biomonitoring Network sampling programme, with the Miramichi River Environmental Assessment Committee in New Brunswick. While these partnerships were formed out of necessity and the desire for mutual survival, they have empowered NAACAP and enabled far more successful outcomes than could have been accomplished independently.

An integral part of NAACAP's role in the community and one of its greatest successes is its participation on environmental advisory committees. Membership in these committees promotes open communication and dialogue across sectors. The provincial government initiated the Incinerator Road Environmental Committee (IREC) in 2007. IREC creates a positive environment for communication between the industries established and operating near Incinerator Road, community-led environmental interest groups and all three levels of government, to enable cross-sectoral dialogue on environmental issues to protect the watershed areas from industry impacts (Incinerator Road Environmental Committee, 2007). NAACAP's role in IREC is increasingly important as anthropogenic pressures continue to put strain on the watershed despite efforts from industry to minimize impact under the direction of the committee.

In addition to NAACAP's participation in IREC, it has a representative on the Environmental Advisory Committee of St. John's. This committee "provides information and advice to the St. John's Municipal Council on environmental issues affecting the City ... or the community" (City of St. John's Environmental Advisory Committee, 2012, p. 1). As of 2015, the committee is made up of two representatives of the city council, six city residents and a representative of each of the following non-profit organizations: Nature Conservancy Canada, Food Security Network of Newfoundland and Labrador, NAACAP and the Newfoundland and Labrador Environmental Industry Association. NAACAP's participation in municipal advisory committees gives means and opportunity to provide guidance on municipal by-laws and critical issues relating to development, sedimentation and wetland protection, to name a few. Through this role, NAACAP is a critical communication link between the city and the community on environmental matters.

NAACAP's research projects involve collaboration with other organizations in the community. Academic institutions in the region, including Memorial University of Newfoundland and its affiliate Fisheries and Marine Institute, are invaluable partners in many research initiatives, providing critical expertise and insight into project development and implementation. Other institutions, including the College of the North Atlantic and St. Mary's University (Nova Scotia), have also provided instrumental assistance through various means, including intern support, field equipment, and training. A number of research projects have also been made possible through collaboration with government agencies, such as the Water Resources Management Division within the province's Department of Environment and Conservation.

The relationships with these institutions have not only enabled NAACAP to conduct extensive research in the region, they have also increased the credibility of the results locally, at a governmental level and within the scientific community.

Partnership with other regional, community based non-profits is a core component of NAACAP's initiatives in the region. NAACAP partners with numerous environmental NGOs, non-profits and community groups, such as Rotary of Waterford Valley, Kelligrews Ecological Enhancement Program and the Newfoundland and Labrador Environmental Industry Association. NAACAP is also a member of the Newfoundland and Labrador Environment Network, a provincial non-profit organization whose aims are to facilitate public education on the environment. Our partnerships help disseminate information to a larger audience in the Northeast Avalon region and naturally extend to other regions of the province and Atlantic Canada.

NAACAP is a small non-profit with two staff. Without the support of external organizations and volunteers, NAACAP would not have the manpower for the field work for many projects. The interest and dedication of these organizations has enabled NAACAP not only to complete projects but also to extend the length and scope of many projects, including water quality monitoring across the region. Furthermore, NAACAP's relationship with diverse non-profits eliminates the 'silo effect' and provides the organization with new perspectives on environmental issues in the region, thus improving the effectiveness of the application of IWM in the region.

In summary, NAACAP's success in forming partnerships throughout the community provides the foundation for IWM in the Northeast Avalon region. With input, buy-in and direction from academia, government, environmental organizations and other NGOs, NAACAP enables positive change and improves capacities for sustainable conservation in the region. By working directly with all stakeholders in watershed management, and the communities themselves, a better understanding of the key issues in these unique watersheds and increased awareness of how to implement effective conservation and rehabilitation efforts are achieved.

Challenges and lessons learned

On an organizational level, NAACAP has faced many challenges in its 22-year history, most predominantly the procurement of stable, core operational funding. On average, the NAACAP`s annual operating budget has been CAD 90,000. When the NAACAP was formed in 1993, Environment Canada's ACAP directly funded the core operating budget. In 1997 the ACAP funding transitioned into a project-based model. ACAP was renamed the Atlantic Ecosystem Initiative in the mid-2000s. This reflected the federal government's desire to encourage more comprehensive ecosystem-based projects that included collaborations among the organizations in Atlantic Canada rather than individual watershed-based projects (Environment Canada, 2015). Although the focus of the funding shifted, ACAP groups throughout Atlantic Canada continued to be the main groups funded through AEI. Thus, NAACAP continued to receive most of its funding primarily from Environment Canada, with occasional project-related support from provincial government departments. In 2014, the funding eligibility restrictions under AEI were changed, and there were greater opportunities to fund Atlantic-based NGOs; coalitions and networks of organizations; research and academic institutions; and Aboriginal governments and organizations. With greater demands for funds and no increase in the total budget, it is unclear whether NAACAP and other organizations that have become reliant on Environment Canada for ongoing support will see a continuation of historical patterns or a decline.

The funding challenge has limited NAACAP's implementation of IWM in two ways. First, it has limited the organization's ability to undertake long-term planning and initiatives. NAACAP has learned that a project-based funding approach promotes a focus on short-term project timelines and limits the ability to track the long-term trends, impacts and outcomes of projects. NAACAP's previous projects have attempted to understand and address the issues and dynamics of the watersheds in the region but have been limited by the duration of the funding.

Second, funding challenges have impacted the focus areas of NAACAP's projects. NAACAP's activities are restricted to eligible funding activities as determined by Environment Canada. NAACAP's reliance on federal funding has resulted in a vulnerability and dependence on an increasingly competitive grant agency. This is illustrated by NAACAP's water quality monitoring activities in the region. Due to funding restrictions, NAACAP's past monitoring has been piecemeal, limiting an accurate assessment of what is happening in the region and therefore limiting initiatives to improve conditions. It has been driven more by the funder's interests than by the best interests of the region. Water quality monitoring has taken place in many projects, such as wetland surveys and watershed health assessments, but data have not been synthesized or analyzed due to limited capacity. However, throughout the 2015 strategic planning process, the NAACAP identified priority areas where project funding will be directed, providing more complete data collection and analysis (NAACAP, 2015).

While funding-diversification discussions have been considerable in recent years, no targeted action has been taken towards procuring alternative funding sources to complement the project funding available from the government. However, in 2015 NAACAP completed a strategic plan. The strategic planning session was undertaken by the board of directors and staff, with the support of the provincial Department of Business, Tourism, Culture and Rural Development. Prior to the session, feedback on the priorities and activities of NAACAP was sought from internal and external stakeholders by means of a survey. Based on the feedback received, the NAACAP team identified four priority areas to guide the organization for the upcoming five years: funding diversification; urban streams; municipal collaboration; and the development of a formal outreach plan to increase community engagement. Funding diversification includes fundraising and sponsorship endeavours, and opportunities that have previously not been undertaken or investigated.

Another organizational challenge the NAACAP has addressed is greater diversity in directors. As a non-profit, the NAACAP is run by a volunteer board, whose members are mostly local science researchers. While this has significant benefits in terms of research quality and technical expertise, it has also contributed to the vulnerability of the organization in recent years. Lack of diversity limits the understanding of the factors affecting the watersheds in the region. NAACAP has recognized the importance of diversification in terms of sampling alternative points of view but also to diversify skill sets and maximize the work of the organization, especially in relation to outreach and educational initiatives. NAACAP aims to include more representatives from the private sector, including those specializing in public relations and fundraising.

Finally, implementing IMW in a region with 78 watersheds in 15 municipalities has inherent complications, most notably in effectively managing watersheds that lie across multiple

municipalities and jurisdictions. It is an ongoing challenge and priority to encourage municipalities to implement proactive, consistent policies and programmes and to recognize the
impact of their actions on the watershed as a whole. NAACAP works with each of the municipalities to provide guidance and facilitate cross-municipal dialogue and collaboration.

NAACAP recognizes the importance of IWM and the lessons of the past 22 years; it continues to take important steps to more effectively incorporate the elements of IWM into the
work conducted.

Future endeavours

The opportunities and the need for conservation in the Northeast Avalon region are endless.
Anthropogenic changes in the region continue to present new challenges and threats to
the health and sustainability of watersheds. NAACAP aims to continue to address these
challenges using an IWM approach: local initiatives and involvement; partnerships and collaboration; watershed basis; and aquatic health.

Long-term water quality monitoring throughout the region will continue to be completed
by NAACAP and its partners. NAACAP is entering the second year of a three-year water
quality monitoring project at sites on each major river in the Northeast Avalon region to get
a complete geographical picture of what is happening throughout the region. This project,
Water Quality Monitoring of Regional Rivers, presents the opportunity to understand long-
term changes throughout most of the watersheds in the region, identify and understand
trends, and complete a more detailed analysis. Filling in the data gaps will provide policy
and decision makers a better understanding of the current state of the aquatic environments
in their jurisdictions. It is hoped that decision makers will be able to make more informed
decisions regarding the development and implementation of environmental policy. Also,
other stakeholders in the region will have a better understanding of the impact of human
activity and other stressors on aquatic environments. Alongside rigorous long-term water
quality monitoring, NAACAP will also continue to enable members of the community to
conduct independent water quality monitoring through coordinating training and capacity
building. This will promote additional stewardship in the region.

Another upcoming project is Riparian Zone Remediation in the Rennie's River Watershed.
The project will utilize existing data collected by NAACAP in the Rennie's River watershed
to identify key areas in which to implement soft solutions, like native species planting and
reach clean-ups.

Conclusion

NAACAP is continuing to grow and expand partnerships and coalitions with organizations
inside and outside the Northeast Avalon region. Networking and sharing ideas has always
been a cornerstone of NAACAP's activities and will continue to be so, fostering new and
innovative approaches to watershed management. NAACAP will continue to work with all
levels of government to coordinate and synchronize conservation efforts and environmental
policy to improve consistency and clarity on environmental issues in the 15 municipalities.
NAACAP is committed to continued dialogue with government to improve and streamline
existing policies, especially in the area of incident reporting and enforcement.

IWM is a complex process, especially in a region with numerous, diverse watersheds, such as the Northeast Avalon region. With the assistance of partners, government and other stakeholders, NAACAP has and will continue to explore and overcome challenges, engage with the community, and protect the watersheds of the region for the betterment of all.

Acknowledgements

Many people were helpful in the making of this article. Thanks to Jen Daniels, Phoebe Metcalfe, Hershal Pandya, Bob Helleur, Corinna Favaro, Brad Strang, Bill Stoyles and Diana Baird. Without support from these people, this article would not have been possible.

Disclosure statement

No potential conflict of interest was reported by the author.

References

Bartellas, E. (2010). *Freshwater joys handbook*. Retrieved from http://www.naacap.ca/site/?page_id=1269

City of St. John's. (2003). *St. John's harbour clean-up: Phase 2*. Retrieved from http://www.ndal.com/userfiles/image/HarbourCleanupReportP2Feb2003.pdf

City of St. John's Environmental Advisory Committee. (2012). *Terms of reference* (Unpublished).

Environment Canada. (1985). *Canada water act*. Retrieved from http://laws-lois.justice.gc.ca/eng/acts/c-11/page-1.html

Environment Canada. (1998). *ACAP: Keeping up with communities*. Environment Canada, Catalogue No: ISSN 1481-305X

Environment Canada. (2003). *The Atlantic Coastal Action Program (ACAP) - Celebrating the successes of long-term community partnerships*. Catalogue No: En4-31/2003E, ISBN: 0-662-34472-3.

Environment Canada. (2015). *Funding programs*. Environment Canada. Retrieved from http://www.ec.gc.ca/financement-funding/default.asp?lang=En&n=923047A0-1#_11

Environment Canada. (n.d.). *ACAP communities in action*: Brochure.

Favaro, C. (2014). *Workshop proceedings: Working together for responsible erosion and sediment control*. Retrieved from http://www.naacap.ca/site/wp-content/uploads/2015/01/Sediment-Workshop-Proceedings.pdf

Ficken, D. (2006). *Regional watershed survey: Nut brook drainage basin, St. John's - stream analysis of a river system in a local industrial zone*. Retrieved from http://www.naacap.ca/site/wp-content/uploads/2011/03/nut_brook_report.pdf

Gardner Pinfold Consulting Economists. (2005). *Atlantic coastal action program results-Based management and accountability framework logic model*. Halifax, Nova Scotia: Gardner Pinfold Consulting Economists.

Government of Canada. (1985). *Fisheries Act*. Retrieved from http://laws-lois.justice.gc.ca/eng/acts/f-14/page-1.html

Government of Canada. (1990). *Canada's green plan: Canada's green plan for a healthy environment*. Cat. No. En21-94/1990E, ISBN 0-662-18291-W. Retrieved from http://cfs.nrcan.gc.ca/pubwarehouse/pdfs/24604.pdf

Incinerator Road Environmental Committee. (2007). *Terms of reference*. Incinerator Road Environmental Committee.

Northeast Avalon Atlantic Canada Action Plan. (n.d.). Reports. NAACAP. Retrieved from http://www.naacap.ca/site/?page_id=709

Northeast Avalon Atlantic Canada Action Plan. (2009). *Freshwater joys*: NAACAP. Retrieved from http://www.naacap.ca/site/wp-content/uploads/2015/07/Freshwater-Joys.pdf

Northeast Avalon Atlantic Canada Action Plan. (2013). *Comprehensive environmental management plan*: NAACAP. Retrieved from http://www.naacap.ca/site/?page_id=1563

Northeast Avalon Atlantic Canada Action Plan. (2015). *Conservation action annual report: 2014-2015*: NAACAP. Retrieved from http://www.naacap.ca/site/wp-content/uploads/2015/09/NAACAP-Annual-Report-2014-15.pdf

St. John's Harbour ACAP. (1997). *Comprehensive environment management plan*. Author.

Statistics Canada. (2011). *Focus on geography series, 2011 census*. Retrieved from http://www12.statcan.gc.ca/census-recensement/2011/as-sa/fogs-spg/Facts-csd-eng.cfm?LANG=Eng&GK=CSD&GC=1001519

The Telegram. (2013a, August 3). 'I wouldn't swim in it'. *The Telegram*. Retrieved from http://www.thetelegram.com/News/Local/2013-08-03/article-3337870/I-wouldnt-swim-in-it/1

The Telegram. (2013b, August 3). On Our Radar: Examining the fate of the St. John's harbour bubble. *The Telegram*. Retrieved from http://www.gulfnews.ca/News/Local/2013-12-14/article-3543154/On-Our-Radar%3A-Examining-the-fate-of-the-St.-John%26rsquo%3Bs-harbour-bubble/1

The Telegram. (2015, May 18). Harbour bubble gone after more than a century. *The Telegram*. Retrieved from http://www.thetelegram.com/News/Local/2015-05-18/article-4150539/Harbour-bubble-gone-after-more-than-a-century/1

Implementing integrated watershed management in Quebec: examples from the Saint John River Watershed Organization

Marie-Claude Leclerc and Michel Grégoire

ABSTRACT

Water management in the province of Quebec has evolved rapidly in recent years. Public consultation led the provincial government to adopt a Quebec Water Policy in 2002, which was reinforced with the passing of the Quebec Water Act in 2009. This legislative tool enabled the creation of 40 watershed organizations responsible for implementing integrated watershed management (IWM). This article explains the context in which IWM has evolved in the province of Quebec. It also describes the successes, challenges and lessons learned by the Saint John River Watershed Organization in implementing IWM in a transboundary watershed.

Introduction: Quebec, land of water

The province of Quebec has a total area of 1,667,712 km². The total area covered with water is 366,896 km², or 22%. Of this amount, nearly 13% is freshwater (Québec, 2014). Quebec contains 4500 rivers, 430 major watersheds and half a million lakes. One hundred watersheds have a drainage area of over 4000 km², and 30 lakes have an area of over 250 km² (Québec, 2002). Integrated watershed management (IWM) in Quebec is performed in 40 defined hydrographic areas by watershed organizations (WOs). WOs in Quebec are mandated for implementing the governance of water under the Act to Affirm the Collective Nature of Water Resources and Reinforce Their Protection, adopted in 2009 and commonly known as the Quebec Water Act. In accordance with this act, each WO is mandated to develop and update a Water Master Plan with various stakeholders, and promote and monitor its implementation.

If Quebec is characterized by its abundance of freshwater, it can also be described by its forests, which cover more than 50% of the province (MFFP, 2015). One hundred and forty-five hydropower plants in 30 different watersheds produce 198.9 TWh annually (MERN, 2015). Quebec's soil is fertile ground for 30,000 farms covering an area of 60,000 km² and generating annual revenues estimated at CAD 1.1 billion in 2014 (StatCan, 2014). Tourism ultimately characterizes this province, with its distinct culture, diverse landscapes and changing seasons. It attracts over 90 million tourists annually and generates CAD 11 billion in tourism revenue (MT, 2015).

Background

In this section, the evolution of water legislation and IWM in Quebec is outlined. As described in subsequent sections, IWM in Quebec consists of a mix of long-standing supporting legal traditions, such as common heritage, and a strong belief that water is a public good to be managed by government. In this context, more recent principles of IWM include the watershed as the management unit, bottom-up planning characterized by strong engagement with stakeholders, and water master planning based on sound technical information, with annual financial support from the Quebec government.

In 1968, the government of Quebec, through the Legendre Commission, raised questions concerning the legal status of water to ensure its accessibility for all. Since that time, new water laws, which are discussed in subsequent sections of this article, have resulted in a significant paradigm shift. Water, with its new collective status, is separate from land, thereby entrusting water management and control to the state (MRN, 1975).

It was in 2002, with the adoption of the Quebec Water Policy (Québec, 2002), that the recommendations of the Legendre Commission report (MRN, 1975) and the subsequent Beauchamp Commission report (BAPE, 2000) began to have a real impact on water management in Quebec. In fact, the three main recommendations of the Beauchamp Commission report were implemented following Quebec provincial Environmental Public Hearings Bureau recommendations. The recommendations which emerged from that consultation and introduced three key foundations of IWM in Quebec at the time were to (1) adopt a provincial water policy; (2) adopt a law for water and aquatic environments; and (3) establish a royalty system for water withdrawals and discharges.

The first recommendation was implemented in 2002 with the adoption of the Quebec Water Policy (Québec, 2002). Although adopted unanimously by the parliament, the policy initially did not reform the principal mechanism for water governance, as the position for a State Minister for Water Resources, which would oversee the integration of government decisions related to water, was abolished by a government change after only five months between 2002 and 2003. As explained in subsequent sections, the Quebec Water Act is fundamental to the planning and implementation of IWM because it provides for the establishment and funding of WOs, and requires them to work with other stakeholders (government, private, and non-government) to complete a Water Master Plan (background to the Quebec Water Act is available at http://www.mddelcc.gouv.qc.ca/eau/politique/index-en.htm).

Since 2002, IWM has been gradually implemented throughout southern Quebec, initially through the funding and designation of 33 watershed groups for prioritized and specific watersheds. A total of CAD 85,000 was provided to each WO, an initial CAD 20,000 for creation and CAD 65,000 annually for operations. The practice of IWM was expanded in 2009 with the restructuring of the previous 33 watershed management zones' geographical boundaries and the creation of seven new IWM zones and a corresponding WO, each receiving an initial CAD 50,000 for start-up. The result was a total of 40 IWM zones that completely cover southern Quebec, each with its own WO. Each of the 40 WOs now receives annual funding of CAD 123,500 for operations. A Federation of Quebec Watershed Organizations, known in French as the Regroupement des organismes de bassins versants du Québec (ROBVQ, https://robvq.qc.ca/), has been created. Based in Quebec City, its mission is to share experiences and work with the individual WOs to promote water governance to enhance the achievement of IWM and sustainable development. All 40 WOs are members of ROBVQ, even though there is no obligation to join and an annual cost of CAD 500.

The second recommendation of the Beauchamp Commission report (BAPE, 2000) was to adopt an Act for Water and Aquatic Environments. This was partially achieved in 2009 with the passing of Bill 27, the Quebec Water Act (Quebec Official Publisher, 2009). This legislation replaced the Water Benefit Act, and integrates the previous set of sectoral water laws. The Quebec Water Act enabled WOs to be officially mandated and recognized by the government. This was an important step to facilitate more effective water governance and collaboration with all interested parties to develop Water Master Plans. However, this law does not define commitments or goals. Instead, it provides broad principles of governance of watersheds as well as the management of the Saint Lawrence River. It defines regulations for water withdrawals, and above all, water as a common heritage. Finally, the last recommendation of the report was implemented in a system of royalty charges for large users, framed by the Regulation on Royalties Payable for Water Use adopted in 2010 under the Environmental Quality Act (http://legisquebec.gouv.qc.ca/en/ShowDoc/cr/Q-2,%20r.%2010).

The Quebec Water Policy (2002)

The Quebec Water Policy stands apart from other jurisdictions, e.g. the US Clean Water Action Plan (United States Environmental Protection Agency, n.d.) and the EU Water Framework Directive (European Commission, 2016), most notably by its bottom-up approach. The Quebec Water Policy endorses participatory governance and IWM. It authorizes local organizations, lake and river associations, regional environmental councils and concerned citizens to create WOs throughout southern Quebec. Water stakeholders participate in round tables convened by WOs and decision-making processes regarding the overall priorities for action for the hydrological zones defined and characterized in the Water Master Plans.

Principles of the Quebec Water Policy

Sasseville and Maranda (2000, p. 32) define IWM as "an ongoing and interactive process based on participatory governance and consultation of all stakeholders (politicians, economic and community-based sectors) for comprehensive planning and standardization of protective measures and the use of the ecosystem's resources in a sustainable manner". The five main priorities of the Quebec Water Policy, which guides the practice of IWM, are:

1. Reform water governance.
2. Implement integrated management for the Saint Lawrence River basin.
3. Protect water quality and aquatic ecosystems.
4. Pursue wastewater treatment and improve water services management.
5. Promote recreational activities related to water. (Québec, 2002)

The reform of water governance relies on local and regional leadership from stakeholders, along with leadership from the provincial government, coordination and accountability for actions, and ultimately the empowerment of stakeholders (Québec, 2002). These elements are the pillars of the Quebec model. The five lines of action related to the reform of water governance are:

a. Revise the water jurisdictional framework.
b. Implement IWM.

c. Develop water knowledge.
d. Implement economic instruments for water governance.
e. Reinforce Quebec's partnerships and relations.

The Quebec Water Policy affirms that one of its major intervention priorities is IWM and states, "this management mode, characterized by a territorial approach that defines the watershed as the management unit for waterbodies, offers the solution with the most advantages to sectoral management of water" (Québec, 2002, p. 17). In fact, the Beauchamp Commission report states that the watershed or catchment will be the administrative unit for water planning and management in Quebec, with leadership from local and regional stakeholders. As mentioned, this idea is reflected in the policy, which also stipulates that all decisions on land use in relation to water use should be made on the basis of the watershed rather than on municipal boundaries. After more than 10 years of implementation, this policy has led to many successes, which are described below. However, there are shortcomings, and not all water-use decisions have been made using the watershed as the management unit.

The report also states that in Quebec, water is to be managed in an integrated and non-sectoral manner, which implies a global vision shared by all government bodies rather than the responsibility of a single isolated department, varying according to the skills of the decision-making bodies (mainly government), as in the past. This creates opportunities for greater coordination, accountability, efficiency and effectiveness of actions. In contrast to the United States, where the Environmental Protection Agency adopted an institutional approach that demarcates all aspects of water management, water governance in Quebec is assigned to the WO, whose role is to ensure overall coordination across its watershed. The 2009 Quebec Water Act specified the mission of these organizations. Article 14 states:

> The Minister can …, for each of the hydrologic units …, enable the creation of a body whose mission is to develop and update a water master plan and facilitate and monitor its implementation, ensuring balanced representation, within that body, of users and of stakeholders from such sectors as the government, Native, municipal, economic, environmental, agricultural and community sectors.

The final cornerstone of the Quebec Water Policy is the empowerment of stakeholders – a significant paradigm shift. Raîche (2005) explains that voluntary involvement of the stakeholders (governmental or not) will be favoured in accordance with the regulations. All stakeholders will be considered responsible and will need to take responsibility. So all users, polluters or not, have the right to be heard. He added that the acceptance of the user-pays and polluter-pays principles promoted in the Quebec Water Policy requires the right to be heard, or at least warrants it. The Quebec approach is distinctive in its participatory and non-binding nature, in contrast to the approach adopted in France, where land-use plans and water management (in French, *schémas d'aménagement et de gestion des eaux*) and master plans for water management (*schémas directeurs d'aménagement et de gestion des eaux*) are binding on unified territorial schemes (*schémas de coherence territorial*) and implemented with funding from water agencies that collect income taxes and fees (Eaufrance, n.d.). Referred to here as 'water governance', the development and the implementation of plans in Quebec is to be decided by all stakeholders and based on the clear founding principles of transparency of information and citizen consultation. This voluntary approach also means that the implementation of actions is based on good faith and the ability of stakeholders to finance those actions.

Practical application of the policy for WOs

As noted previously, between 2002 and 2009, 33 IWM groups were created as non-profit organizations, covering 25% of southern Quebec. These organizations were established by local communities in response to the principles of the Quebec Water Policy to modernize water management. The IWM groups, whose territories were defined by the government, managed primary watershed boundaries contributing to the Saint Lawrence, Ottawa or Saguenay Rivers. Other watershed management groups were also formed, even though they were not recognized or funded by the government. If the Quebec Water Act of 2009 had not brought an end to this situation by defining hydrological management zones, which included sometimes several watersheds, 450 agencies responsible for water management could have been created, resulting not only in financial disaster but also in a problem of standardizing planning tools.

The WOs are independent and autonomous entities. The members of each WO define its mission and elect its directors (by electoral divisions or votes by the more or less 60 members). The government has the discretion whether to recognize a WO, depending on the development of a Water Master Plan. The WOs therefore have significant similarities but also differences in their focus and in the way they operate. For example, some WOs have a board of directors, which makes recommendations related to the Water Master Plan in addition to providing corporate management oversight. Other WOs have a separate board that develops the Water Master Plan based on consultative round tables. These round tables can be defined according to a given specific watershed, such as the Conseil de gouvernance de l'eau des bassins versants de la rivière Saint-François (Round Table for the Saint-François River Watershed), or by sector, for example the Corporation d'aménagement de la rivière l'Assomption (Round Table of Planning Agencies in the l'Assomption River Watershed).

To ensure scientific validity, WOs have been responsible for preparing water management plans with the help of officials in various provincial ministries, the municipalities, and other experts since 2002. The first generation of Water Master Plans were completed with few resources and little support, and with so much freedom there were many innovations as well as abject failures. In 2014, the second generation of Water Master Plans were filed with the Ministry of Sustainable Development, Environment and the Fight against Climate Change (MDDELCC) for analysis. Guidelines have gradually been drawn up with the help of ROBVQ, mandated by the ministry to support the WOs in writing their Water Master Plans. A table of contents and guidelines were provided to WOs, which they have been encouraged to use. At the time of writing, training and guidance have also been provided to most WOs to enable them to better understand the use of the Water Master Plan as a management tool and their role in its design. In addition, the WOs have had access to data from 21 departments and agencies of the Québec government and contributors to ACRI-Geo (Networking Co-operational approach for Geographic Information). This gives them high-quality socio-economic (e.g. demographic, economic) and biophysical (e.g. water quality, water quantity, soil, land use) data derived from rigorous methodologies for use by their scientific committees (about 75% of WOs have a scientific committee). Moreover, as these data are updated regularly, the WOs will be able to annually update their Water Master Plans with as much rigor as for their initial plan, as agreed upon with the MDDELCC.

In 2015, the Water Master Plans are well understood by all WOs. While those directly involved with WOs are familiar with Water Master Plans, there is a need to provide continuing education to other stakeholders and the general public, the need and scope of which are still poorly understood.

The Quebec Water Act

The 2009 Quebec Water Act provides interesting elements that were lacking in the 2002 policy. In particular, it confirms the legal status of water as a "common heritage of the Quebec nation" and forbids the use of market rules (Québec, 2002, p. 9), which also excludes it from the North American Free Trade Agreement. (Halley & Gagnon, 2009, p. 42) state that the concept of "common heritage" refers to the idea that water should be seen as an "inheritance received from previous generations that must be preserved and passed on to future generations, necessitating rational and sustainable management and exploitation". Thus, water is treated as an intergenerational and public good in Quebec.

The Quebec Water Act also makes the state the guardian of water, and as such it should legislate to conserve water quality and quantity in the general interest and for sustainable development. In doing so, as explained by Halley and Gagnon (2009), the act establishes the opportunity to receive compensation for the degradation of water quality or quantity, regardless of fault. Enacted exclusively by Quebec's attorney general, the Water Act authorizes compensation in the name of the state – "guardian of the interests of the nation in matters of water resources", the restoration of sites, compensatory work or compensation for damage caused to water resources. This provision is consistent with remedies that already existed in Canadian common law. Drawing on what is already in place in Canada, the law also has given to holders of the "common heritage" a statutory recourse in water protection, authorizing a tribunal to control non-compliance with the act.

For several reasons, the Quebec Water Act is very important to the WOs. The WOs are and will remain non-profit organizations with autonomous status. Nevertheless, as emphasized by Raîche (2005, p. 2), "Their legitimacy is … increased in some way by the possibility of Ministerial recognition, provided they meet the mandate set by law and by Ministry of Sustainable Development, Environment and the Fight against Climate Change financial agreements." In addition to ensuring the neutrality of the decision makers who are the members of a WO, the law establishes, as already stated, that water management in Quebec is organized according to watersheds. Moreover, Article 15 states: "The Minister must also send a copy of the plan to government departments and agencies as well as regional municipalities, metropolitan communities and local municipalities whose territory is included in whole or in part in a hydrographic unit covered by the plan, so that they take it into account when exercising powers conferred to them by law in the domain of water or in any other domain affecting water." This encourages municipalities and regional municipalities to include the elements of the Water Master Plan that apply to their jurisdiction through municipal initiatives, such as land-use planning and development plans. As for municipalities and regional municipalities taking the Water Master Plan developed by WOs into consideration, several WOs, such as those in the Charlevoix-Montmorency, Sainte-Anne and l'Assomption River watersheds, have made agreements with their regional municipality that ensure alignment between the Water Master Plans and land-use and development plans. This alignment is made possible by coordinating the two planning initiatives early in the drafting of both documents.

There has been progress in developing partnerships between municipalities and regional municipalities, for which the standardization of planning tools is not necessarily as important as management or leadership. For example, during the development of the Water Master Plan of the De la Capitale watershed, stakeholders coordinated their implementation and have undertaken over 400 actions in support of the plan.

The Quebec Water Act established a new licensing regime for water withdrawals. Halley and Gagnon (2009) maintain that it is very different from the previous approach prescribed in the Environmental Quality Act. While the authorization of the minister's power is discretionary, it must now take into account a set of new criteria, including: effects of climate change; rules for sustainable management, water equitability and efficiency; the precautionary principle; the hierarchy of uses (prioritizing the need for drinking water); and balancing the needs of ecosystems, economy and society. Surprisingly, in the Environment Quality Act, the minister is accorded the power to assess an application for a certificate of authorization not just in environmental terms but also in social and economic terms. Even more unusual, the government (Cabinet) has bound itself for the first time to these criteria, which are the same as those applicable to the minister's decision, when water withdrawals are subject to a process of assessment and examination of environmental impacts. The Water Withdrawal and Protection Regulation was approved and initiated in 2014 in order to complete the implementation of the Quebec Water Act (Quebec, n.d.1). These criteria reflect important elements of contemporary IWM in Quebec.

Finally, the Quebec Water Act defines a method for royalties for water resources to be based on different rates for different purposes. The Regulation on Royalties Payable for Water Use was adopted in 2010 and is based on voluntary reporting of the first withdrawal that uses more than 75,000 litres of water daily, in accordance with the Regulations on Reporting Water Withdrawals adopted in 2009 (Quebec, n.d.2).

Practical applications of the Quebec Water Act for WOs

In 2009, the passage of the Quebec Water Act also created challenges for some aspects of water management in the province. For example, the geographic jurisdictions of the majority of existing river basin committees were greatly enlarged, in one case from 800 km^2 to 26,000 km^2. In some other cases, two organizations were merged and therefore had to change their organizational structure to meet the new requirements for the financing of a single entity. This situation occurred when the Fouquette River Basin Organization and the Kamouraska River Basin Organization merged to become the Kamouraska, L'Islet, Rivière-du-Loup Watershed Organization. A final challenge occurred when seven new WOs were created in hydrologic areas where none had previously existed, such as the Organisme de bassin versant du Témiscamingue, Organisme de bassins versants Duplessis, and Organisme de bassin versant du fleuve Saint-Jean (Figure 1).

Currently, there are several factors that constrain IWM in the province. These include: (1) lack of legal power for the implementation of Water Master Plans; (2) complexity of territorial divisions (i.e. watershed boundaries versus municipal boundaries creates an obstacle for harmonizing planning tools); (3) lack of consistency in defining territorial limits; (4) difficulty in effectively involving elected representatives; and (5) differing priorities between those responsible for water governance and those responsible for land management.

Financing of IWM in Quebec

Funding for IWM is a challenge in Quebec because IWM is so broadly defined. Several financial appraisals have been made by MDDELCC. Although it represents a major investment for

1 - Organisme de bassin versant Abitibi-Jamésie
2 - Organisme de bassin versant du Témiscamingue
3 - Agence de bassin versant des 7
4 - Comité du bassin versant de la rivière du Lièvre
5 - Organisme de bassins versants des rivières Rouge, Petite Nation et Saumon
6 - Organisme de bassin versant de la rivière du Nord
7 - Conseil du bassin versant de la région de Vaudreuil-Soulanges
8 - Conseil des bassins versants des Mille-Îles
9 - Corporation de l'Aménagement de la Rivière l'Assomption
10 - Organisme des bassins versants de la Zone Bayonne
11 - Association pour la gestion intégrée de la rivière Maskinongé
12 - Organisme de bassins versants des rivières du Loup et des Yamachiche
13 - Bassin Versant Saint-Maurice
14 - Société d'aménagement et de mise en valeur du bassin de la Batiscan
15 - Organisme des bassins versants des rivières Sainte-Anne, Portneuf, La Chevrotière
16 - Organisme des bassins versants de la Capitale
17 - Corporation du bassin de la Jacques-Cartier
18 - Organisme de bassins versants Charlevoix Montmorency
19 - Organisme de bassin versant du Saguenay
20 - Organisme de bassin versant Lac-Saint-Jean
21 - Organisme des bassins versants de la Haute-Côte-Nord
22 - Organisme de bassins versants Manicouagan
23 - Organisme de bassins versants Duplessis

24 - Société de conservation et d'aménagement du bassin de la rivière Chateauguay
25 - Comité de concertation et de valorisation du bassin de la rivière Richelieu
26 - Organisme de bassin versant de la baie Missisquoi
27 - Organisme de bassin versant de la Yamaska
28 - Conseil de gouvernance de l'eau des bassins versants de la rivière Saint-François
29 - Organisme de concertation pour l'eau des bassins versants de la rivière Nicolet
30 - Groupe de concertation des bassins versants de la zone Bécancour
31 - Organisme de bassins versants de la zone du Chêne
32 - Comité de bassin de la rivière Chaudière
33 - Conseil de bassin de la rivière Etchemin
34 - Organisme des Bassins Versants de la Côte-du-Sud
35 - Organisme de bassins versants Kamouraska, L'Islet, Rivière-du-Loup
36 - Organisme des bassin versant du Fleuve Saint-Jean
37 - Organisme des bassins versants du nord-est du Bas-Saint-Laurent
38 - Organisme de bassin versant Matapédia-Restigouche
39 - Conseil de l'Eau Gaspésie Sud
40 - Conseil de l'eau du Nord de la Gaspésie

Figure 1. The integrated water management zones of the 40 watershed organizations of the province of Quebec.

Quebec, in the domain of water it is not part of an overall implementation package for IWM but an investment made independently from the new arrangements for water governance.

Funding for each WO is provided to fulfil the mandate given by the Quebec Water Act. This mandate is, as mentioned before, to revise and develop a Water Master Plan, and to facilitate and monitor its implementation. This core funding must be distinguished from the financing of measures and actions required to implement the Water Master Plan. Together, the core funding and the funding for implementation of the plans constitute IWM financing.

Although the situation has improved since 2002, funding for IWM in Quebec is still modest and continues to be a challenge. As noted earlier, between 2002 and 2009, funding levels were CAD 65,000 per year for each of the 33 river basin committees. The 2009 redeployment of IWM led to an increase in funding to CAD 123,500 per year for each of the current 40 WOs. This represents an investment of CAD 5,000,000 annually for the entire province, representing CAD 0.63 per person per year. Funding is used exclusively for permanent staff, office space, telecommunications, committee meetings and other expenses necessary for the operation of the Water Master Plans.

Water governance in Quebec

At present, there is no provincial funding for actions arising from the Water Master Plans, and no formal implementation strategy. As a result, implementation varies among WOs. Examples of activities include: enforcing and monitoring wastewater treatment facilities; educating farmers on the importance of implementing good practices on their fields; monitoring the quality of surface waters; undertaking an inventory of and developing management plans for wetlands; prioritizing and restoring degraded aquatic habitats; reducing the area of bare soil surfaces in flood zones; setting up irrigation schemes; and implementing riparian tree and shrub planting programmes.

While funding programmes for the implementation of Water Master Plans are not available, WOs can access other funding programmes for specific activities. Two examples are the provincial Green Premium for agricultural activities, which provides funding for addressing a variety of agricultural issues, and the federal government's Partnerships for Recreational Fisheries Conservation Program, which provides funds to restore, rebuild and rehabilitate recreational fisheries habitat. The fact that there is no explicit funding programme for undertaking Water Master Plan actions causes financial difficulties, with stakeholders often paying expenses from their own pocket. It is also difficult for the WOs to monitor implementation as they are often not directly involved in the funding or implementation of activities associated with the Water Master Plan.

For ROBVQ, the solution for effective implementation of Water Master Plans is the development of a provincial Action Plan for Integrated Water Resources Management, the realization of which supports the 40 Water Master Plans. This Action Plan should have a five-year strategy and funding programmes from the 11 ministries involved (MDDELCC, Ministry of Municipal Affairs and Land Occupancy, Ministry of Agriculture, Fisheries and Food, the Executive Council Office, Department of Forestry, Wildlife and Parks, Department of Energy and Natural Resources, Department of Public Safety, Ministry of Health and Social Services, Department of Transport, Department of Tourism and Ministry of International Relations and Francophonie). The Action Plan should also: (1) identify provincial priorities based on the management needs of the individual WO; (2) determine management targets; (3) outline the timelines, resources and actions of the various ministries in support of implementation; and (4) define indicators for monitoring progress. Above all, financial programmes should be available to local stakeholders (e.g. municipalities, Aboriginal communities, farmers, foresters, environmental organizations, WOs) so that the actions defined in the Water Master Plan can be better achieved.

Future challenges of the Quebec model

The integrated water management model of Quebec is remarkable, and WOs can begin to develop and implement a long-term participation process that goes beyond the exploratory stages.

However, the model's implementation is challenging, given the significant paradigm shift it represents for municipalities, local interest groups and the provincial government, which no longer plays a sole leadership role. Equity financing, standardization of planning tools, implementation of the Regulation on the Protection of Drinking Water Sources, adoption of a law on wetlands, and revision of the Environmental Quality Act are some factors that must be addressed in order to improve IWM success in Quebec. Despite these challenges, the authors believe that the new water governance model in Quebec has a promising future.

ROBVQ and its members are currently drafting what is termed the Blue Book for Water Management in Quebec to offer a common vision for water (to be posted on the ROBVQ website, https://robvq.qc.ca/). The Organisme de bassin versant du fleuve Saint-Jean, or Saint John River Watershed Organization (SJRWO), is participating in this work and is an excellent example of a WO established by and in its community. To further illustrate the strengths, opportunities and barriers to IWM in Quebec, the SJRWO, one of the most recent WOs, will be presented as a case study.

The Saint John River Watershed Organization

The SJRWO was established in 2010. Its implementation was carried out by an existing local environmental organization, the Témiscouata Lake and Madawaska River Management Society, which manages an interprovincial cycling route between Quebec and New Brunswick (the Petit Témis) and pursues environmental projects in the transboundary (Quebec and New Brunswick) Madawaska River watershed. Since then, both organizations have worked separately and collaborated on specific issues, such as invasive plants.

The SJRWO, like all of Quebec's WOs, is an independent non-profit organization. It has a board of 22 directors, and 4 staff. Located in Témiscouata-sur-le-lac (Figure 2), a municipality of 5000 inhabitants, it focuses on implementing a Water Master Plan developed with water stakeholders for the Quebec portion of the transboundary Saint John River basin.

The Saint John River watershed as a whole

Most WOs in Quebec are responsible for managing watersheds that discharge into the Saint Lawrence River within provincial borders. However, because political borders rarely follow watershed boundaries, there are some exceptions. On Quebec's south-eastern slopes of the Appalachian range are the headwaters of one of the greatest watersheds of Atlantic North America, the Saint John River basin. Originally named Wolastoq, the river was occupied by the Maliseet First Nations when renowned French explorers Pierre Dugas de Mons and Samuel de Champlain discovered the region in their search for a site that would become the first French colony in North America. In 1604, they named the river "fleuve Saint-Jean", or Saint John River, on the celebration day of Saint John the Baptist, 24 June. Two days later, 26 June, the explorers established themselves on Saint Croix Island further south, and then began the French occupation in North America.

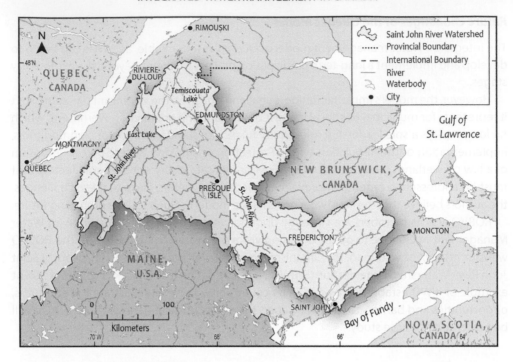

Figure 2. The Saint John River Basin.

The Wolastoq watershed corresponds almost completely to the Maliseet Nation's ancestral land. More recently, during the nineteenth century, the 55,000 km² river basin was divided between the American state of Maine and the Canadian provinces of New Brunswick and Quebec, which respectively contain 37%, 50% and 13% of the surface area of the watershed (Figure 2). This multi-jurisdictional watershed is now influenced by the different cultures and languages of two countries and one First Nation, making an integrated approach to watershed management even more challenging.

With a population of more than 500,000 and an average of 10 people per km², the whole basin is very rural, with a low population density and an aging and diminishing population in small communities (Kidd, Curry, & Munkittrick, 2011). With 70,000 inhabitants, the city of Saint John, New Brunswick, on the Bay of Fundy, is the largest urban centre in the basin. Today, Maliseet communities are found in eight locations, including one in Quebec, six in New Brunswick and one in Maine.

The forestry industry is a major economic activity in the basin. In addition, the productive central plains of the watershed, in New Brunswick and Maine, have become prominent potato production areas for North America.

Until the seventeenth century, the basin was practically untouched by human transformation or influence and was almost completely covered with old-growth Acadian forests. Today, there are rare forest parcels that have never been harvested in the watershed. Rapid colonization from the 1880s to the 1950s, and industrialization from the 1920s to the 1970s, generated changes and pollution issues along the river. An International Joint Commission (IJC, 1954) report from 1954 describes the great storage and hydroelectric production infrastructure potential of the basin. Some of the projects explored have been realized, such as the Beechwood Dam, which affect the natural flow regime of the Saint John River. The report frames these large infrastructure projects as "water resources development". Pollution issues were mostly linked to poor waste disposal practices and the discharge of raw waste water and contaminants.

Another International Joint Commission report, from 1977, states that "the waste discharge from food processing industries which rely on large quantities of water for production, the discharge of municipal wastes from the larger urban centres, and serious soil erosion problems in the potato producing areas are principal concerns in basin-wide water quality management" (IJC, 1977, p. 19). During the 1970s and 1980s, through the implementation of various regulations and programmes, such as the Quebec Environmental Quality Act adopted in 1972 (CanLII, n.d.) and the Quebec Water Treatment Program in 1978, significant pollution problems were incrementally addressed (MDDEFP, 2012). Furthermore, the agriculture and forestry sectors began to consider water quality impacts in their management practices. For instance, in the Quebec portion of the basin, the Groupement forestier de Témiscouata, a forest landowners cooperative, was one of the first in Quebec to obtain Forest Stewardship Council certification. Since 2012, nearly all cattle farms have sealed facilities to store manure and reduce contaminants in agricultural runoff. As stated in the Saint John River State of the Environment Report (Kidd et al., 2011, p. 159), "Since the last extensive surveys in the 1960s and 1970s, there have been improvements in various indicators of environmental quality along the Saint John River."

The Saint John River is one of the few basins in Quebec that does not discharge into the Saint Lawrence River. In fact, it flows south eastward out of the province, giving the basin more of an Atlantic Canada identity than a central Canadian one. This reality is both exciting and challenging. The upstream position of the Quebec portion confers a strategic responsibility upon upstream users towards their downstream neighbours, where numerous Maine and New Brunswick communities draw river water for their domestic water supply.

For the remainder of this article, 'the Saint John River watershed' will refer only to the portion located in Quebec, part of the upper Saint John, representing 55 municipalities throughout an area of 7213 km². The upper Saint John is the portion of the watershed that is upstream from Grand Falls, a 23-metre drop of the river. It represents nearly one-third of the entire watershed.

Principal threats to the Saint John River watershed

The Saint John River watershed hosts 18 lakes larger than 1 km², with natural landscapes, tranquillity and wilderness characteristics that attract more and more people every year. Témiscouata Lake, 40 km long, covering an area of 67 km², and with a depth of 75 metres, is the most significant lake, with more than 1000 residences directly on its shoreline. All of the lakes play an important role in the regional identity. These once-isolated, crystal-clear salmonid habitats are experiencing nutrient inputs from various point and non-point sources, such as construction sites, road ditches, forestry and agricultural runoff. This results in the expansion of macrophytes, algae and periphyton, and exacerbates eutrophication in lakes. These contaminants are slowly making what were once extraordinary lakeside sites less attractive. Excess aquatic biomass can exacerbate hypoxia in the hypolimnion. One such documented case in Long Lake seems to demonstrate that the lake trout population is being put at risk by a reduction of oxygen, resulting in habitat loss (OBVD, 2014).

Intense nutrient inputs continue largely due to the lack of appropriate erosion control practices. This excessive erosion comes mostly from anthropogenic activities such as municipal road and ditch construction and maintenance. Even though standards to ensure efficient runoff controls exist (MTQ, 2011), they are rarely applied. The SJRWO has focused its resources on providing training on erosion control techniques for road maintenance workers and foremen. The SJRWO is looking forward to continuing to work with municipalities on the adoption of environmental ditch management plans to realize effective erosion control in the drainage network.

The Saint John watershed, like most others in Quebec, faces the increasing threat of invasive species. This recent challenge appears to be under control now for some species but still requires great attention, education and rapid and costly mobilization. The common reed (*Phragmites australis*) appeared in the Saint John watershed adjacent to the Trans-Canada Highway in about 2002 and is now spreading rapidly, causing losses in biodiversity. The SJRWO has worked with the Ministry of Transportation to map plant colonies, but eradication has yet to be done to avoid downstream propagation. Another plant that appeared in the watershed during the same period is the giant hogweed. The lack of action regarding invasive plant species in the watershed that are not directly harmful to humans is one example of the consequences of having no funding to implement the Water Master Plan. Another example of a non-native species in the watershed is the muskellunge (*esox muskellunge*). Native to other parts of Quebec, the muskellunge is a ferocious predator fish at the very top of the freshwater food chain in North America. It was intentionally introduced in the 1970s to Lake Frontier, a small upstream lake of the basin. More than 40 years later, the fish has colonized most of the Saint John River and its main tributaries, creating problems for wildlife fishery managers and salmonid anglers. Muskellunge has had a major impact on the native fish community of the basin. Impassable structures such as the Allagash Falls (in Maine) and the Madawaska Dam (in Edmundston, New Brunswick) appear to be the only obstacles to the spread of this species. The SJRWO works on documenting its distribution and facilitating discussion among regional and local actors towards coherent and valuable fishery management of the species.

As in many other watersheds of North America, the transport of fossil fuels by rail close to waterways has increased dramatically. Specific to this region, pressure is rising for an extensive pipeline infrastructure project which would cross the entire Saint John River watershed, from its source to its mouth at the Bay of Fundy. The increase in the transportation of hazardous products comes with a proportionate level of risk of incidents resulting in the contamination of the ecosystem. More than 5000 people obtain their drinking water from Témiscouata Lake, which is just 17 km downstream from the crossing of the proposed Energy East pipeline project under the Cabano River. The local population and decision makers have received no analysis of the risk that the proposed pipeline poses to the river system. The SJRWO produced an informative map with data to highlight the potential risks to water and aquatic ecosystems. This information has been disseminated to elected representatives and the general public to inform the decision-making process and increase understanding of the geography of the project to identify where preventive measures are most needed should the project proceed (http://obvfleuvestjean.com/non-classe/carte-le-transport-de-petrole-*chez*-nous-et-les-risques-pour-leau/).

Inefficient municipal and industrial water treatment infrastructure cause some localized water quality problems. Wastewater plants built in the 1960s do not provide sufficient treatment of wastewater effluent, which is discharged into the river system. Financing infrastructure upgrades can be difficult for rural communities facing rising expenses, aging populations and low incomes. The SJRWO demonstrates the impact of underperforming infrastructure on water quality through taking water samples and having them analyzed. It also collaborates with local municipalities to help them overcome obstacles to updating their water treatment facilities.

Governance structure

The SJRWO is governed by a cross-sectoral board of directors. There are 22 directors, including six municipal mayors; two representatives each from the agricultural, forestry and citizenry sectors; and one representative each from the lake association, wildlife conservation, environment conservation and economic sectors. There is also one seat reserved for First Nations. Finally, there are five advisory (non-voting) seats for ministry representatives. The proportions of this diverse representation are described in the Reference Framework for Watershed Organizations (MDDEFP, 2012) and required by the Quebec Water Act. The voting seats require 20–40% representation from citizen groups, 20–40% representation from the economic sector and 20–40% representation from municipalities.

The board members represent a region with a length of 400 km. This distance makes regular face-to-face administrative meetings impossible. This limitation makes the infrequent meetings (four board meetings annually) susceptible to overloaded agendas and low interest in the various complex issues that need to be considered. To address this challenge, the staff at the SJRWO strive for extremely well-prepared agendas, with meetings limited to two hours. Another crucial approach to meetings is to ensure that all decision makers understand problems both close to and far from their own locations, as well as learning about each other's specific expectations and goals. Representatives in designated seats, such as the regional municipality, are not provided a travel allowance. However, representatives in elected seats such as citizens or lake associations have their travel expenses reimbursed by the SJRWO. In 2015, the director (chief of staff) initiated an annual personal visit with each board member to facilitate a better understanding of the respective realities and sensitive issues of each. Within the board, the awareness of being part of the same river system is growing steadily. Many who previously believed they had nothing in common now sit at the same 'round table' talking about the health of the watershed they share. The current president of the SJRWO, Yves Marquis, is well respected in the region and often cited as a model in the agro-environmental community. As a family-farm Jersey cow milk producer, following in the footsteps of his father, Yves is committed to community engagement.

Since most volunteer board members have busy professional or vocational and community lives, meetings of the SJRWO need to be well structured to ensure that important issues are addressed promptly and decisions made. Accommodating the wide-ranging interests of the various board members is a challenge that defines the true meaning and intent of the expression 'integrated' watershed management.

A smaller watershed (3000–4000 km^2) would permit more frequent meetings and more in-depth conversations to address the diversity of opinion on issues. Despite the challenges of distance, it is the opinion of the authors that SJRWO has interested and engaged board members committed to respectfully sharing various point of views.

Water Master Plan

The Water Master Plan was developed following the guidelines established by ROBVQ (n.d.) and the *Water Master Plan Development Guide* (Gangbazo, 2011), a manual prepared by MDDELCC for the use of the WOs. The manual leads staff through the development of an overview, a diagnosis and an action plan for the identified watershed. The stakeholders need to be consulted, and a science review committee needs to be set up. The science committee can take diverse forms depending on the availability of scientists close to the basin or knowledge of the basin. The Water Master Plan for the Saint John River was mostly developed by the SJRWO staff, with input from the board of directors, citizens and stakeholders, which was sought through various public consultation mechanisms such as focus groups and public meetings. Public meetings were very productive and elicited open discussion among participants. Developing the Water Master Plan took three years. The plan has been reviewed by and comments received from eight ministries involved with water issues. This process took a year and a half to complete and was supervised by the Integrated Water Management Service of the Water Policies Division of MDDELCC. The Water Master Plan of the Saint John River basin has received final approval from the minister and is publicly available on the SJRWO website (http://obvfleuvestjean.com/plan-directeur-de-leau).

The Water Master Plan will need to be revised annually, with a substantive and validated update every five years. Actions identified in the Water Master Plan are either actions that need to be taken or actions that a stakeholder agrees to do. The implementation of actions depends generally on the will of stakeholders or the availability of funding. The progress of the action plan is measured annually by a follow-up and evaluation programme based on agreed performance indicators.

Citizen's empowerment: a key to long-term success of IWM

While the main branch of the Saint John River does not flow through the province of Quebec, some of its major tributaries do. That poses challenges for citizen participation and the sense of belonging to the watershed. Why would someone bother with the fate of a river that does not flow in his or her own province or country? Individuals are attached to their local lake or river – the one they can see and feel. Few are aware of how their 'special place' is connected to a larger basin. Since the IWM approach is very recent in Quebec and fragmented by political boundaries, a sense of belonging to and an overall vision for the Saint John River watershed has not yet emerged. The journey towards strong and permanent citizen engagement and empowerment is long, and it is only with continuous communication efforts that this sense of belonging can be achieved.

Most of the work of SJRWO relates to long-term objectives and complex issues which are not easy for the public to grasp. For this reason, many activities need to be tangible and show visible results in order for the public to understand the benefits of IWM. For example, the SJRWO recently managed to eliminate the goldfish population in a golf club pond by drying it up. This population was threatening to reach Témiscouata Lake. Once reported by the media, this action caught the attention of many, contributing to the recognition of the role of SJRWO and raising awareness of the issue of invasive fish introductions. Another action is the publication of a short, accessible and simple guide to health for the lakes (http://obvfleuvestjean.com/lacs/). This serves to spur discussion on lake health among residents and raise stakeholder awareness of emerging environmental issues.

WO staff need to be strong communicators and advocates for IWM. Staff are viewed as environmental ambassadors upon which the public depends to advocate on their behalf. Public expectations are high, and WOs are expected to facilitate solutions to existing and emerging water and ecosystem problems in areas of large geographic size, complex eco-systems and several political jurisdictions. A major challenge for WOs is to find the optimum balance across the various initiatives that their role requires. WO staff keep being pulled between large-scale actions, such as supporting lake associations in their endeavours to advocate for stronger regulations within municipal jurisdictions, and small-scale ones, such as working directly on riparian restoration for a specific site.

Conclusion

The implementation of IWM in the Quebec portion of the Saint John River basin is recent. The Water Master Plan submitted to the province of Quebec just received final approval in August 2016. This Water Master Plan details the objectives and action plan identified for implementation by various partners over the next five years.

In comparison with most other WOs in southern Quebec, the SJRWO is in an enviable position in that it is dealing with management issues where the landscape has experi-enced minor change. For the most part, the focus is on rejuvenating good-quality eco-systems, rather than restoring highly altered ones. This is an opportunity not many other parts of the world have. The SJRWO staff have worked diligently to document the water-shed attributes, understand the basin's conditions, and tie the various stakeholders into a 'water web' for communicating information and knowledge. With the Water Master Plan having just received final approval, many stakeholders are ready to get involved in advancing the following proposed central vision for the Quebec portion of the water-shed:'In the Saint John River watershed, the maintenance of integral ecosystems, source of excellent water quality, constitutes the base of a legacy built on healthy transboundary relations' (OBVFSJ, 2015).

The creation of the SJRWO in 2010 occurred just before the publication of *The Saint John River: A State of the Environment Report* by the Canadian Rivers Institute in 2011 (Kidd et al., 2011). In 2012, the World Wildlife Fund Canada hired an advisor for the Saint John River through its Living Rivers Initiative Program. Subsequently, in 2013, the Saint John River was recognized as a Canadian Heritage River. In 2014, a watershed planning initiative was begun by the Maliseet Nation Conservation Council, and a freshwater health assessment for the Saint John basin was released (WWF, 2014). This momentum has permitted the SJRWO to rapidly connect with other agencies and organizations and create a communications web across the whole transboundary basin. Much is happening, and there is general consensus that an all-embracing approach is needed for the Saint John River. Much work remains. The SJRWO, the voice for the Quebec portion of the basin, intends to make a significant contri-bution towards a healthy Saint John River ecosystem.

Disclosure statement

No potential conflict of interest was reported by the authors.

References

BAPE. (2000). Rapport de la commission sur la gestion de l'eau au Québec. [Report of the commission on the management of water in Quebec]. *Bureau d'audiences publiques sur l'environnement*, 286. Retrieved from http://www.bape.gouv.qc.ca/sections/archives/eau/rapport.htm

CanLII. (n.d.). Environment quality act, CQLR c Q-2. Retrieved from http://www.canlii.org/en/qc/laws/stat/rsq-c-q-2/latest/rsq-c-q-2.html

Eaufrance. (n.d.). Schémas d'aménagement et de gestion des eaux (SAGE) [Water management schemes]. Retrieved from http://www.eaufrance.fr/s-informer/agir-et-participer/planifier-et-programmer/?id_article=86

European Commission. (2016). The EU water framework directive - Integrated river basin management for Europe. Retrieved from http://ec.europa.eu/environment/water/water-framework/index_en.html

Gangbazo, G. (2011). Guide pour l'élaboration d'un plan directeur de l'eau: un manuel pour assister les organismes de bassin versant du Québec dans la planification de la gestion intégrée des ressources en eau [Guide for the development of a water master plan: A manual to assist watershed organizations of Quebec with integrated water management]. *Québec. Ministère du Développement durable, de l'Environnement et des Parcs*, 329. Retrieved from http://www.mddelcc.gouv.qc.ca/eau/bassinversant/guide-elaboration-pde.pdf

Halley, P., & Gagnon, C. (2009). Le droit nouveau de l'eau au Québec [The new water right of Quebec]. *Gaiapresses*. Retrieved juillet 3, 2009, from http://gaiapresse.ca/analyses/le-droit-nouveau-de-leau-au-quebec-109.html

IJC. (1954). Interim report to the governments of the United States and Canada on the water resources of the Saint-John River basin, Quebec, Maine and New-Brunswick. *International Joint Commission*. Retrieved from http://www.ijc.org/files/dockets/Docket%2063/Docket%2063%20Conservation%20&%20Regulation%20Final%20Report%201954-01-27.pdf

IJC. (1977). Water quality in the Saint John River basin. *International Joint Commission*. Retrieved from http://www.ijc.org/files/dockets/Docket%2096/Docket%2096%20Report%20to%20Gov..pdf

Kidd, S. D., Curry, R. A., & Munkittrick, K. R. (2011). *The Saint John River a state of the environment report* (p. 175). Fredericton, New Brunswick: Canadian Rivers Institute, University of New Brunswick. Retrieved from http://www.unb.ca/research/institutes/cri/_resources/pdfs/criday2011/cri_sjr_soe_final.pdf

MDDEFP. (2012). Gestion intégrée des ressources en eau: cadre de référence [Integrated water resources management: Terms of reference]. *Ministère du Développement durable, de l'Environnement, de la Faune et des Parcs* (p. 36). Québec. Retrieved from http://www.mddelcc.gouv.qc.ca/eau/bassinversant/GIRE-cadre-reference.pdf

MERN. (2015). Statistiques énergétiques en ligne [Energy statistics online]. Ministère de l'Énergie et des ressources naturelles. Retrieved from http://www.mern.gouv.qc.ca/energie/statistiques/statistiques-production-electricite.jsp

MFFP. (2015). Ressources et industries forestières, Portrait statistique édition 2015 [Resource and forest industry statistics, 2015 edition]. Retrieved from http://www.mffp.gouv.qc.ca/publications/forets/connaissances/portrait-statistique-2015.pdf

MRN. (1975). ministère des Richesses naturelles, Commission d'étude des problèmes juridiques de l'eau. *Rapport de la commission d'étude des problèmes juridiques de l'eau* [Report of the committee for the study of the legal problems of water]. ministère des Richesses naturelles, Commission d'étude des problèmes juridiques de l'eau.

MT. (2015). *Tourisme en chiffres* [Tourism statistics]. (p. 2). Tourisme Québec: Ministère du tourisme. Retrieved from http://www.tourisme.gouv.qc.ca/publications/media/document/etudes-statistiques/tourisme-chiffres-2014.pdf

MTQ. (2011). Méthode du tiers inférieur pour l'entretien des fossés routiers, Guide d'information à l'intention des gestionnaires des réseaux routiers [Method of maintenance for the bottom third of road ditches; information guide for managers of road networks]. *Ministère des transports du Québec*, 15. Retrieved from http://www.bv.transports.gouv.qc.ca/mono/1079063.pdf

OBVD. (2014). Étude diagnostic du lac Long [Study diagnosis of long lake]. *Organisme de bassin versant Duplessis*, 74. Retrieved from http://obvfleuvestjean.com/wp-content/uploads/2015/08/%C3%89tude-lac-Long-FINALE.pdf

OBVFSJ. (2015). Plan directeur de l'eau du bassin versant du fleuve Saint Jean [Saint John River basin Water Master Plan]. Organisme de bassin versant du fleuve Saint-Jean, 461. Retrieved from http://obvfleuvestjean.com/plan-directeur-de-leau

Quebec. (n.d.1). Water law. Retrieved from http://www.mddelcc.gouv.qc.ca/eau/politique/index-en.htm

Quebec. (n.d.2). Laws and regulations. Retrieved from http://www.mddelcc.gouv.qc.ca/publications/lois-reglem-en.htm

Québec. (2002). Politique nationale de l'eau [Quebec water policy]. Gouvernement du Québec, 103. Retrieved from http://www.mddelcc.gouv.qc.ca/eau/politique/politique-integral.pdf

Québec. (2014). Le Québec chiffres en main – Édition 2014 [Quebec figures in hand – 2014 Edition]. Institut de la statistique du Québec, 72. Retrieved from http://www.stat.gouv.qc.ca/quebec-chiffre-main/pdf/qcm2014_fr.pdf

Quebec Official Publisher. (2009). Bill 27 (2009, chapter 21) an act to affirm the collective nature of water resources and provide for increased water resource protection. Retrieved from http://www2.publicationsduquebec.gouv.qc.ca/dynamicSearch/telecharge.php?type=5&file=2009C21A.PDF

Raîche, J-P. (2005). Organisme de bassin versant: gestion ou gouvernance: management or governance [Watershed organizations: Management or governance]. Vecteur environnement, 38, 11–13.

ROBVQ. (n.d.). Boîte à outils Plan directeur de l'eau (PDE) [Water master plan toolbox]. Retrieved from https://robvq.qc.ca/guides/pde

Sasseville, J. L., & Maranda, Y. (2000). L'administration publique de l'eau par bassin versant [The public management of water]. Vecteur environnement, 33, 32–42.

StatCan. (2014). Statistics Canada. Government of Canada. Online consultation. Retrieved from http://www.statcan.gc.ca/tables-tableaux/sum-som/l02/cst01/agri02e-fra.htm

United States Environmental Protection Agency. (n.d.). About the Clean Water Act (CWA) Action Plan. Retrieved from https://www.epa.gov/compliance/about-clean-water-act-cwa-action-plan

WWF Canada. (2014). Fresh water health assessment for the Saint John River. World Wildlife Fund. Retrieved from http://awsassets.wwf.ca/downloads/sjr_health_assessment_wscda_230514.pdf

Setting the stage for IWRM: the case of the upper Kiskatinaw River, British Columbia

Reg C. Whiten

ABSTRACT

Integrated Water Resources Management has gained prominence in British Columbia due to growing concerns about drinking water quality and supply and risks from cumulative land-use and climate change. Experiences are examined in the upper Kiskatinaw River of the Peace River basin and the source watershed for the city of Dawson Creek. Though there is no formal decision-making capacity, efforts led by the city have focused on balancing intensive resource industry development activity with its stewardship objectives despite not having a formal role in watershed governance. Through investments in planning, characterization and monitoring, the city is well positioned now to further achieve facilitating interest-based solutions.

Introduction

The US Army Corps of Engineers constructed the city of Dawson Creek's water supply system during World War II. It would have been hard to imagine just how much the landscape in the upper Kiskatinaw River watershed in north-east British Columbia would change in the nearly 70 years that followed (Figure 1). The unstable silty drainage system that gave definition to its original Woodland Cree name (*kîskatinâw sipi*, steep hill or cutbank river) is known for its very erodible riparian terrain, with high natural spikes in turbidity after spring freshets and intense rainfall. Very little land-use activity in those days would have added to this impact in the watershed. Other contemporary water management challenges relate to incidental surface water diversion and sediment loading in the Kiskatinaw River from expanding gas-industry roads and pipeline infrastructure.

Traditional resource use by the region's Aboriginal people of Treaty 8 (BC) and rural settlers recorded plentiful harvests of ungulates (e.g., moose, deer and caribou) and fish (e.g., rainbow trout, Arctic grayling, whitefish, dolly varden, pike and pickerel). Today, much of the Aboriginal use has been curtailed in this watershed due to increased cumulative land-use change. This has led to the degradation of some sub-basins due to riparian habitat loss, sedimentation and species displacement (Forest Practices Board [FPB], 2011a). Some previously harvested

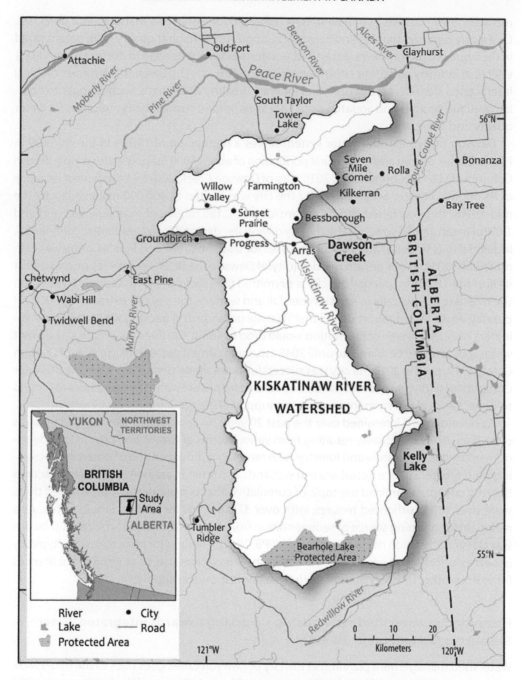

Figure 1. The Kiskatinaw River watershed in the Upper Peace River Basin, British Columbia.

species, such as boreal caribou, are listed as at risk in the region and considered key indicators of hydro-ecosystem health.

Other Aboriginal communities based at Kelly Lake (Cree Nation, First Nation and Apetokosan Nation, Kelley Lake Métis Settlement Society) have relied on the watershed as part of their traditional use territory for hunting, fishing and trapping. However, there is no record of their interests ever having been assessed for watershed management purposes. One study reported that key indicators of caribou winter habitat quality have deteriorated in the upper Kiskatinaw River watershed, and further work is needed to document the direct effects of industry and other contributing factors (FPB, 2011b). Given that hydro-ecological interactions are critical elements for maintaining healthy watersheds, such changes to riparian, wetland habitats and aquatic species are all important sustainability indicators for water quality and flows.

Dawson Creek's existing water system serves a population of 12,115 in the city, 689 in Pouce Coupe, plus an additional rural population of about 3000 for bulk water supply (British Columbia Oil & Gas Commission, 2016; City of Dawson Creek, 2007). At the time of issuance of its original licence, water supply for the city was limited to 400,000 gallons/day. By the mid-1990s, this was raised to 3 M gallons/day, or 0.183 m^3/sec of river flow demand and 1.8% of mean annual flow. In 2014, the city daily water demand for *all* residential, commercial, industrial (gas fracking) and agricultural uses for Dawson Creek and Pouce Coupe was estimated at 550 litres per person per day (City of Dawson Creek, 2014). This represents about 44% of the 18,000 m^3 per day the city is permitted to draw from the Kiskatinaw River. When the river is low – typically in late summer, fall and winter – the city can extract only 9000 m^3 per day, and a recent report stated that 'If [shale gas industry] fracking water use was eliminated, daily per-person consumption would drop to about 435 litres. As a result, the current water source would be adequate until 2048 assuming current patterns of use' (City of Dawson Creek, 2014, p. 4). In 2013 the city was successful in obtaining permission from the BC Water Comptroller to divert up to half its licensed volume for private water transport sales to Alberta to generate local government revenues to support shale gas industry development.

As development intensified over the past 20 years, so too have concerns about impacts on surface flows and quality resulting from various forms of Crown and private land development by gas, agriculture and forestry, with resulting rapid expansion of watercourse crossings and greater surface disturbance of wetlands and riparian areas. A 2011 Forest Practices Board study that examined the topic of cumulative effects management determined there were over 1200 authorized tenures, with over 37 crossings located on erodible soils. As a result, this continuing source of sedimentation from human activities and some sub-basins was classified as a high risk to water quality (FPB, 2011b). Tenures for resource use are typically issued as (short-term) permits, as with most shale gas water use, or as (long-term) licences for various other industry options, including domestic water supply.

Planning for watershed stewardship – working towards Integrated Water Resources Management (IWRM)

Watershed management planning at the city of Dawson Creek goes back to the mid-1980s, leading to an Integrated Watershed Management Plan in 1991, one of the first for a municipal water purveyor in the province (Government of British Columbia, 1991). The purpose was to detail 'a land and resource management plan for the watershed that would ensure that water quality, quantity and timing of flows is given the highest priority in all resource management decisions affecting domestic drinking water supply, forestry, fish and wildlife habitat, recreation, oil and gas, mining and other land use activities'. While the work fit well with the evolving practice of IWRM, it was not defined in the terms of the internationally accepted definition, which flowed from the 1992 World Summit on Sustainable Development and was expressed by the Global Water Partnership as 'a process which promotes the co-ordinated development and management of water, land and related resources, in order to maximize the resultant economic and social welfare in an equitable manner without compromising the sustainability of vital ecosystems'.

For various reasons elaborated on in this article, there have been obstacles to enabling the necessary shared decision making to enable IWRM, but considerable progress has been made. While the intent may have existed to enable shared decision making in the 1991 Integrated Watershed Management Plan, it was essentially an exercise in high-level strategic planning. The necessary institutional arrangements and supporting legislation were not put in place, including designated roles and mandates for relevant ministry staff. After 20 years, several of these barriers are targeted as priorities by the province through creation of a watershed management governance structure. It remains to be seen whether the necessary financial capacity and decision-making authority are put in place for enabling this function beyond what the city has been able to undertake on its own. Such an agenda for action has been presented at the provincial level by independent public-interest bodies (Brandes, O'Riordan, & Simms, 2017).

Critical issues in those early days centred around flow availability to support city water demand. That process, led by the BC Ministry of Forests, involved a full cross-section of watershed stakeholders, which at the time was dominated by concerns with range and forest sectors, with interest in prospective future oil and gas development. At that time, activity was limited to 83 active wells averaging 0.18 ha in size. An important step was taken through this Integrated Watershed Management Plan with the creation of a registered Notation of Interest by the province in the Kiskatinaw River watershed, which ensured that attention was given to water management concerns in all land-use and development referrals. On completion of the Integrated Watershed Management Plan, it was stated that 'there are short-comings and gaps, notably a detailed set of resource management guidelines which set down measures and constraints to be followed by all resource users (to be prepared for future versions of the report)' (Government of British Columbia, 1991, p. 1).

Subsequent land and resource management planning in the mid-1990s, and increased regulatory oversight from passage of the Forest Practices Code, led to watershed assessments and the fostering of ecosystem-based forest harvesting practices. In the 1999 cabinet-approved Dawson Creek Land and Resource Management Plan, the upper Kiskatinaw River watershed was given recognition as a Domestic Water Supply Area, which provided some guidelines for land-use management to protect drinking water supply, but it was not given recognition as a Community Watershed, since that status applies only to watersheds less than 500 km^2 in area. It is likely that had ministerial discretion been used to create the Community Watershed designation, there would have evolved a higher level of oversight in resource development controls to prevent key issues such as induced sedimentation, gas industry waste disposal and wetland degradation.

Notwithstanding a formal request by the city's mayor and council in 2014, the province has not yet accepted the city's request to have it recognized as a Designated Watershed under the BC Oil & Gas Activities Act (Sec. 35) of the Environmental Management and Protection Regulation (British Columbia Oil and Gas Commission, 2016). If such a change were implemented, it would raise the level of collaboration in reviewing development referrals. More importantly, it would highlight issues where some activities might need to be restricted in certain sensitive areas where the impact on water quality was a greater concern, but it would also address many other relevant considerations related to hydro-ecosystem management and protection. The BC Oil & Gas Environmental Management Guideline is an important planning tool because it stipulates that in a Designated Watershed 'it is the responsibility of the applicant to ensure, and demonstrate to the Commission, that an operating area does not cause a material adverse effect on a designated watershed' (British Columbia Oil & Gas Commission, 2016, p. 13).

Without the formal shared decision-making frameworks described above, the city continues to have limited ability to influence water management decision making, relying primarily on incremental site-specific referrals by a land-use proponent. A more formal role would provide advance delineation of sensitive, protected or enhanced management areas related to elevated water risk hazards such as shallow groundwater, or vulnerable riparian and wetland areas, which might impact supply or quality.

First Nations and other Aboriginal groups were not engaged in these earlier watershed-planning processes, but some limited involvement by the Saulteau First Nation did occur as part of the Dawson Creek Land & Resource Management Planning initiative of the mid-1990s. That engagement focused on development of the Bearhole Lake storage reservoir and maintenance of a fish passage as part of the constructed level-control weir. However, Treaty 8 First Nations generally resisted participation in provincial land-use planning because in their view there was inadequate recognition of treaty rights and no articulation of a government-to-government relationship.

This situation has evolved in recent years, as the First Nations engage in different initiatives related to regional cumulative environmental effects and reconciliation agreements, including support for community watershed planning and management objectives. The Saulteau First Nations agreement, for example, has explicit language to address water-quality and flow-management issues but is limited to watersheds with areas of high-priority use and does not specify the upper Kiskatinaw River watershed as an area of community use interest (Saulteau First Nations, 2015). The As'in'i'wa'chi Ni'yaw Nation (also known as Kelly Lake Cree Nation) does claim interest in the upper Kiskatinaw River watershed as part of its Aboriginal rights and title claim under the Comprehensive Land Claim process (Calliou, pers. Comm., 2017). In addition, annual reports are filed for public and First Nations review along with other subsequent consultations relating to the city's water management activities. This impetus has been driven by independent reviews by the FPB (2011a, 2011b) and more recently by the British Columbia Ministry of Health (2011) and the Auditor General of British Columbia (2015). Four of the eight priorities in that latter report recommended that immediate attention be focused on water management related to aquatic ecosystems: watershed condition and risk; low flow and instream flow needs; water quality; and riparian objectives.

In the author's experience, the north-east BC situation as it relates to First Nations engagement differs considerably from the context of modern treaties such as in the Yukon, where explicit terms of engagement and processes are defined with respect to land use and resources management under that territory's Umbrella Final Agreement (1993). In undertaking preparation of a Recommended Regional Land-Use Plan for the Peel River watershed, for example, rigorous efforts were undertaken to apply baseline ecosystem analyses based on rigorous public and First Nations consultations. This was done to determine compatible land uses for collaborative management while applying the precautionary principle to ensure a balance of industrial development activity and traditional resource use (Canadian Environmental Law Association, 2017). The precautionary principle denotes a duty to prevent harm, when it is in our power to do so, even when all the evidence is not available. This principle has been codified in several international treaties to which Canada is a signatory. Domestic law also refers to this principle, but implementation remains limited. The lack of hydro-ecosystem information became one such constraint in defining potential development opportunities for the Peel Watershed Regional Land Use Plan and where protecting wilderness values was part of the overarching statement of intent for that plan. It is significant that this effort resulted in legal challenges and in December 2017, a unanimous ruling of the Supreme Court of Canada was in favour of the appellants, and by extension the work of the Peel Watershed Planning Commission (Supreme Court of Canada, 2017). This precedent-setting decision recognized the need for the Crown to honour both the spirit and intent of the Yukon's 1993 Final Umbrella Agreement and all treaties negotiated in good faith with Canada's indigenous people and their governments. In presenting a framework for co-stewardship decision making in the Recommended and Final Regional Land Use Plans, the Court decision inherently recognizes the importance of fully assessing hydro-ecosystem carrying capacities, biodiversity and traditional resource use values as a foundation for long-term planning.

In the case of the Kiskatinaw Watershed, the Bearhole Lakes Provincial Park and Protected Area were created through the land and resource management planning process to provide permanent protection for the Kiskatinaw River headwaters and other sub-basins. Some additional management direction in the Dawson Creek land and resource management planning provided for a 1000-metre Enhanced Management Zone within the lower Kiskatinaw River main-stem corridor, but no restrictions were established on the types of industrial tenures permitted in that zone (Government of British Columbia, 1998). Some encroachment on this protected management area has occurred in response to management of mountain pine beetle infestations.

By 2003, the city embarked on an update to its watershed management plan with partial funding from the Peace River Watershed Council, and a process of re-engaging all stakeholders in a process of identifying watershed values, and creating a course of action for improved action ranging from education to improved watershed characterization and sediment source control by all land-use sectors (City of Dawson Creek, 2003). That organization operated for the period 2000–2007 with participation of all levels of government, First Nations and stakeholders but lacked sufficient resources and mandate to continue operations.

By this time, it had become clear that risks to water quality and flow from rapidly expanding oil and gas development (e.g., wells, roads, pipelines, waste disposal facilities) was creating increasing uncertainty regarding the upper Kiskatinaw as a viable long-term water source. Other sectors in agriculture (crops and range) also expressed interest in improved watershed management. It became apparent that an interest-based and scientifically sound approach could be pursued for collaborative watershed management to address issues of shared interest. Specific topics that would create a foundation for integrated watershed management focused on the following aspects of desired future outcomes:

- There is a partnership among the stakeholders with the goal to protect the water resource.
- Water quality (after treatment) continues to meet all provincial and federal drinking water requirements.
- Water supply and storage are adequate to meet the demands (both consumptive and non-consumptive).
- Raw water quality is protected from impacts from resource development activities.

- Integrated multiple resource use is compatible with the supply of safe drinking water and the risk of water contamination from all activities is low or moderate in the watershed.
- Integrated and comprehensive plans are developed over time to address all watershed activities and development. (City of Dawson Creek, 2003).

Although both the 1991 and 2003 plans were aimed at promoting IWRM in concept, the approach did not gain the traction or the support needed by the responsible agencies to ensure multiple-industry-sector compliance. However, the work of developing the plans did provide impetus to addressing issues of available water supply and quality protection to support the city of Dawson Creek's needs, as we elaborate in this article. The gap in watershed characterization data, integrated management mechanisms and adequate water legislation that addressed the complex challenges related to both ground and surface water management have all been critical considerations that expanded the conversation. With the increased frequency of several extreme weather events during the period 2010-2016, the inter-related effects of climate change associated with water quality and flooding impacts have gained greater attention from local government, following on from an earlier recommendation about this issue and the need for a more formal approach to IWRM (City of Dawson Creek, 2012; Whiten, 2013).

The upper Kiskatinaw River watershed contrasts with experiences in other jurisdictions in southern BC and elsewhere in Canada, where greater attention been given to the value of a fully applied IWRM approach with the support of senior governments (FitzGibbon, Mitchell, & Veale, 2006; Mitchell & Shrubsole, 2007; Shrubsole, 2017). Conservation Ontario, for example, has provided leadership through many decades of work by Conservation Authorities, where emerging issues and jurisdictional realities are being considered and IWRM is undertaken as a 'process of managing human activities and natural resources on a watershed basis. This approach allows us to protect important water resources, while at the same time addressing critical issues such as the current and future impacts of rapid growth and climate change' (Conservation Ontario, 2017). The challenge remains in trying to move from concept to practice in the upper Kiskatinaw River watershed, and indeed throughout north-east BC.

With elevated levels of parasites, bacteria and pathogens traced to upstream land uses, including ranching, as well as natural sources, the city has devoted its resources to addressing this issue and investments in system upgrades for increasing water storage and expanded levels of water quality treatment (City of Dawson Creek, 2007). This is considered a strategic priority given its limited ability to prevent source water contaminant problems or to influence land-use decisions, and its lack of capacity to address other water resource management issues (Lapp & Whiten, 2012; Whiten, 2012, 2013). Concerns have continued to focus primarily on source water protection in the past decade. Only now, with the promise of water sustainability planning as presented by the province in the BC Water Sustainability Act and Regulations 2016, is the IWRM approach likely to be more formally adopted to build on its leadership efforts in research and monitoring.

The upper Kiskatinaw River watershed experience also provided impetus towards a series of watershed characterization efforts (Government of British Columbia, 2004, 2007, 2008). The city was strongly encouraged by the Northern Health Authority (Regional Drinking Water Team) to undertake development of a source water protection plan in 2006 to bring a clear focus to water quality protection objectives. Funding was provided by the city, with additional financial support from the Peace River Regional District to create a position of a watershed steward to oversee implementation of the city's watershed plans and research programme. Although there are overlapping Aboriginal (Metis and First Nation) interests in the watershed, the source water protection plan process did not undertake any significant participation or follow-up in the planning outputs except, as previously noted, for the recommended water storage infrastructure project at Bearhole Lake. Continued public and First Nations reporting is also required for that reservoir as part of its annual management plan, with a focus on protection of fish and riparian habitat.

Building on best practices to address regional water management challenges

Regional challenges remain concerning many aspects of water supply and flows in the upper Peace basin. As noted earlier, water quality and flow implications have long been known to exist during periods of low flow in drought and in winter. However, it is not yet known how groundwater flows affect the hydrological regime. Shallow groundwater areas, artesian formations and numerous springs in river headwall areas are known to exist at middle elevations in many watersheds and indicate potential risk from development activity. Related to this concern are potential risks to water quality from chemical additives used in hydraulic fracturing, or from surface leakage during or after gas well development. Such issues were identified in various reports leading up to the peak period of activity (2005–2014) for shale gas exploration and development. Significant public and First Nations concerns were raised about potential impacts of shale gas development on drinking water supplies and other related human and ecosystem health considerations (Government of Canada, 2006; Campbell & Horne, 2011; Council of Canadian Academies, 2014; Intrinsik Environmental Science, 2014).

At the regional level, the Montney Water Project was undertaken by the province in partnership with several industry partners and Dawson City. That initiative was aimed at understanding water resources in the major gas field region of the South Peace, including parts of the upper Kiskatinaw River watershed (GeoScience BC, n.d.). Considerable impetus for this work was also drawn from challenges of large-scale hydraulic fracturing operations by Treaty 8 First Nations and rural communities. This public attention led to the provincial government and industry moving to disclose fracturing fluid constituents, develop a new water allocation and use reporting system, improved hydrological modelling and innovative source water or alternative fracking technologies (British Columbia Oil and Gas Commission, 2011; Council of Canadians, 2012; Canadian Association of Petroleum Producers, 2017). A national study by the Council of Canadian Academies (2014) provided further direction towards improved water science research and monitoring related to shale gas development. These various studies highlighted the lack of groundwater information, monitoring and protection and the pressing need for aquifer mapping.

Additional research to characterize basin aquifer profiles using three-dimensional mapping of hydro-geology based on well water pressure gradients, and other groundwater research, has further helped build the picture of shallow and overburden aquifers (GW Solutions, 2012). Another major effort involved airborne electromagnetic mapping in the region, as a cost-effective method of mapping groundwater; the initiative involved extensive First Nations participation and collaboration (GeoScience BC, 2016).

Various interdisciplinary regional water workshops and field tours have helped develop a shared research agenda, building partnerships for best practices projects, and setting new directions for more coordinated research and source water protection implementation (Fraser Basin Council, 2013, Lapp & Whiten, 2012). To close the information gap and improve oversight on development in its water supply area, the city focused its efforts on watershed research and local capacity building. For example, a three-year hydrology study was commissioned with the University of Northern British Columbia to undertake baseline watershed characterization. This study included installation of eight hydrometric stations on the upper Kiskatinaw River to monitor surface and shallow groundwater flows as well as selected water quality parameters (Saha & Lee, 2012). A second study component was aimed at detailed remote sensing analyses to investigate changing land-use patterns (Paul & Lee, 2013). Another collaboration was initiated to improve climate monitoring for protection of wetlands in the watershed. A comprehensive water quality risk assessment focused on oil and gas activities further helped identify potential surface and shallow groundwater contaminant pathways. Together, this research has been increasing the city's ability to implement its watershed management and source water protection plans as a model approach for other rural and First Nations communities. Lessons from this work were recognized by the province of British Columbia in developing its North-East BC Water Strategy.

The city of Dawson Creek's work, combined with other regional water-related research, has demonstrated the potential of locally applied integrated watershed management in the upper Peace basin and particularly where rural, urban and First Nations communities have shared concerns about water security. It has also focused understanding on the weak status of baseline information on surface and ground water quality in the region, which is now being addressed through various strategic stakeholder collaborations in the Northeast Water Strategy (2015). That policy document is aimed at supporting implementation of the province's recent Water Sustainability Act (Government of British Columbia, 2015, 2016), although neither provides explicit recognition of Aboriginal water use rights (InterraPlan, 2014).

Creating a foundation for greater local government involvement in water and land-use decision making provides incentive for considering other water-related environmental issues. For example, public and First Nations concerns about the acceptability of large-scale water diversions for multiple purposes (e.g., agriculture, shale-gas/liquefied natural gas development, or municipal) has not yet been fully examined, but such issues are likely to generate increasing scrutiny related to growing concerns about water availability in the western United States.

With growing water demand in the south-west USA, there is some speculation that existing water diversions, including industrial water pipeline infrastructure, may be considered by international trade law 'commodified resources' under a new or revised North American Free Trade Agreement. Such a legal challenge or negotiated terms by government could enable large-scale water diversions, as envisioned for decades under an updated North American Water and Power Alliance Scheme or other trade agreements (Holms, pers. com., 2016; Lammers, Proussevitch, Frolking, & Grogan, 2013; Nelson, 2017).

Concerns have been raised about all major water transfer infrastructure, be they regional inter-basin projects in the form of existing industry water developments (water pipelines, storage facilities) or trans-basin diversion schemes. These issues include public investment costs, downstream hydro-ecological impacts, commercialization of water diversion through public/private partnerships, and legal questions related to treaty resource use and water rights (Parfitt, 2017a, 2017b).

Reflections on progress towards IWRM

Notwithstanding a major investment by the city in water source protection for the upper Kiskatinaw River watershed, it still operates primarily in a research and monitoring mode and not as a full partner in integrated watershed management, though in a defined decision-making capacity. This is due, in large part, to the current BC regime of deregulation and complaint-based management, in which a system of professional reliance for environmental assessments by industry proponents has shifted the extent of internal regulatory oversight by government regulatory agencies. An extensive review of professional reliance by the University of Victoria's Environmental Law Centre (2015) found that 'professional reliance' was undermining the public interest, given numerous issues related to the rigour of management prescriptions, environmental monitoring and potential conflicts of interest. An earlier report similarly suggested a growing number of major challenges in the professional reliance issue related to riparian protection in terms of public disclosure, system monitoring and reporting (British Columbia Ombudsman Office, 2014).

While drinking water treatment and operations themselves are well supervised through strict operator training standards and oversight by the Northern Health Authority for compliance with the BC Drinking Water Protection Act, source water protection may be more vulnerable on the issue of watershed monitoring and compliance under the current management system. A study on cumulative effects in the south Peace pointed to the need for critical attention to certain key watershed stewardship indicators, such as water use based on over-allocation for the oil/gas sector, lack of reporting of other water use, and the implications of climate change. For example, in certain winter months (November–March), such allocation excesses were reported to range from 95% to 585% in the middle and east portions of the Kiskatinaw River, despite efforts by the regulator to establish a tracking system (the North-East Water Tool) for that purpose. Another measure, referred to as 'riparian intactness' and based on a maximum threshold of 10% incursion on Crown lands, was also being applied in the West Kiskatinaw sub-basin, and is an important indicator given that the province's Riparian Area Regulation for private lands does not apply. Therefore, it is difficult to control sedimentation on inadequately protected lands (Government of British Columbia, 2014).

Serious concerns about the net cumulative effects on water quality based on consistently high key measures, including turbidity, bacteria and parasites, can be linked to riparian degradation. Priority effort must be to continually assess, monitor and mitigate risks to drinking water supply and aquatic ecosystems. Based on the experience in the Kiskatinaw River, affected resource communities such as the city of Dawson Creek are well positioned to advocate for an interest-based approach in achieving these objectives. Constitutionally protected treaty and Aboriginal rights, agreements and treaties must be given full attention, along with other relevant water-related legislation, to identify a compatible and sustainable level of land-use activity.

Numerous examples exist elsewhere in BC and across Canada that demonstrate the value of collaborative watershed stewardship at both local and regional scales (Shrubsole, 2017). Giving priority to drinking water source protection must overcome cross-jurisdictional boundaries and agency mandates to achieve meaningful results. Our experience has shown that land-use planning in BC and elsewhere has recognized the importance of identifying, characterizing and managing risks to drinking water through the widely adopted, multi-barrier approach for source water protection (Environment Canada, 2001). The work is further enhanced by framing watershed stewardship within the context of a hydroscape (Pringle, 2001). Nevertheless, only limited progress has been made in fostering effective operational models of source water protection to consider a greater array of IWRM, including assessments of contaminant risk and hydro-ecological features and interactions. Present challenges lie in trying to foster preventative source water protection objectives within the current regulatory context of results-based management and mitigation frameworks, where development permitting may not fully consider those objectives. Wetland protection, ground–surface water interactions and cumulative land-use change are all key indicators of watershed health and thus sustainability of domestic water supply, yet these components remain poorly understood. Climate change and other landscape perturbations (e.g., drought, fire, mountain pine beetle infestations) are also now providing significant impetus to promote the IWRM approach with emphasis on source water protection.

A growing tension exists with some stakeholders as senior levels of government advance major regional resource projects in the region (e.g., the Site C Hydroelectric Dam, and liquid natural gas water development,[1] which have been unsuccessfully challenged through court actions by First Nations). It is still hoped that further engagement through both legal and political processes will clarify the fiduciary obligations related to land and water resource use for Aboriginal governments on cumulative impact issues and major new resource development projects.

As information gaps in watershed research, policy and regulatory harmonization are being filled, there are prospects for improved decision making on water allocation and protection in north-eastern BC, including the upper Kiskatinaw River watershed. Further impetus is also due to growing public and First Nation concerns about the water quality and supply impacts of key industrial sectors such as mining and shale gas development, which produce contaminant waste by-products from construction and management operations.

Setting the stage for IWRM

Through a combination of watershed characterization, integration of local/traditional knowledge, impact assessments and effective watershed monitoring, it is possible to promote best management practice by all resource users while greatly reducing future water supply and quality risks. With sufficient formal decision-making arrangements, the stage is being set for a more formal approach to IWRM practice, as originally anticipated in the 1991 plan. Although collaboration for IWRM in the wider north-east BC is at an early stage of formal adoption, the strategic application of water resource science is providing a foundation for improved decision making. Through regulatory recognition of the upper Kiskatinaw River as a designated watershed along with a formal watershed management advisory body, for example, the city could demonstrate an effective pilot for watershed co-stewardship for multiple objectives, including gas development. Recent Supreme Court rulings and ongoing legal challenges concerning Aboriginal land and water issues in both regions will also likely provide further impetus for cooperative watershed management (Laidlaw & Passelac-Ross, 2010). With sufficient support and mandate, the creation of water management boards, such as the former Peace River Watershed Council, is now needed to ensure effective dialogue and management oversight at both the basin and sub-basin levels, such as in the upper Kiskatinaw River watershed. At present, it remains unclear how or when meaningful watershed governance will be enabled, but the necessary policy directions, legislative framework and regional watershed baseline research have been created.

Integrated watershed management in its fully applied form, which ensures shared decision making, has yet to be demonstrated in the watersheds of north-east BC. While there is promise in the considerable groundwork of communities, such as the city of Dawson Creek in the upper Kiskatinaw River watershed, integrated watershed management will have a pronounced effect only when properly enforced through appropriate field measures. While further impetus for water sustainability planning is being provided by new provincial initiatives such as the North-East Water Strategy and Water Sustainability Act (2014), there has been inconsistent stakeholder engagement to date. Perhaps the change in provincial government, which occurred in July 2017, will lead to a recognition of the need to refocus efforts on enabling a more accountable and balanced approach to both land-use planning and watershed management practice. For rural and urban communities, as well as First Nations, it is hoped that the renewed focus on a watershed approach in recent years will improve decision making, with a more balanced consideration of groundwater management, instream flow allocations and drinking water quality, while enabling sustainable water use for industrial purposes.

Note

1. BC Hydro's Site C Clean Energy Project has long been proposed as a third dam and hydroelectric generating station on the Peace River in north-east BC. It is now in its second year of construction. It was referred by the BC government in July 2017 to an independent review by the BC Utilities Commission. If completed, it would provide 1100 MW of capacity, and produce about 5100 GWh of electricity each year. Liquefied natural gas projects were a strategic economic development initiative of the former BC government, and following on recent investment decisions of some proponents to withdraw, these projects remain questionable as to market viability and environmental appropriateness for climate change adaptation.

Disclosure statement

No potential conflict of interest was reported by the author.

References

Auditor General of British Columbia. (2015). *Managing the cumulative effects of natural resource development in BC*. Victoria, BC: Office of the Auditor General.

Brandes, O., O'Riordan, J., & Simms, R. (2017). *A revitalized water agenda for British Columbia's circular economy* (Report of the POLIS water sustainability project). Victoria, BC: University of Victoria.

British Columbia Ministry of Health. (2011). *Progress on the action plan for safe drinking water in British Columbia* (Report of the provincial health officer). Victoria, BC: Crown Publications.

British Columbia Oil and Gas Commission. (2011). *Quarterly report on short-term water approvals and use, 2011*. Fort St John, BC: British Columbia Oil and Gas Commission. Retrieved July 11, 2017 from http://www.bcogc.ca/node/6046/download?documentID=1180

British Columbia Oil and Gas Commission. (2016). *Environmental protection and management guideline* (Revised June, 2016). Retrieved July 12, 2017 from http://www.bcogc.ca/node/5899/download

British Columbia Ombudsman Office. (2014). *Striking the balance: The challenges of using a professional reliance model in environmental protection – British Columbia's riparian area regulation* (Public Report No. 50). Retrieved July 12, 2017, from https://www.bcombudsperson.ca/sites/default/files/Public%20 Report%20No%20-%2050%20Striking%20a%20Balance.pdf

Calliou, Clifford. (2017). Personnel communications. Retrieved August 2, 2017, from http://www. kellylakecreenation.com/klcnclaim.htm

Campbell, K., & Horne, M. (2011). *Shale gas in British Columbia: Risks to BC's wate resources*. Drayton Valley, Alberta: Pembina Institute. Retrieved March 31, 2017, from http://www.pembina.org/pub/2263

Canadian Association of Petroleum Producers. (2017). Responsible development – Water. Retrieved July 15, 2017, from http://www.capp.ca/responsible-development/water

Canadian Environmental Law Association. (2017). The precautionary principal. Retrieved from http:// www.cela.ca/collections/pollution/precautionary-principle

City of Dawson Creek. (2003). *Kiskatinaw River watershed management plan*. Prepared by Dobson Engineering Ltd and Urban Systems Ltd. Retrieved July 12, 2017, from http://www.dawsoncreek. ca/wordpress/wp-content/uploads/2011/08/KiskatinawWMP2003.pdf

City of Dawson Creek. (2007). *Kiskatinaw river water source protection plan*. Kelowna, BC: Dobson Engineering Ltd. Retrieved July 12, 2017, from http://www.dawsoncreek.ca/wordpress/wp-content/uploads/2011/10/WatershedManagementPlan.pdf

City of Dawson Creek. (2012). Climate change implications for the City of Dawson Creek and considerations for risk mitigation. Technical briefing document prepared by Reg Whiten to City of Dawson Creek Mayor & Council. Retrieved July 8, 2017, from http://www.dawsoncreek.ca/wordpress/wp-content/uploads/watershed/Climate-Change-Implications-for-DCk.pdf

City of Dawson Creek. (2014). *Sure water campaign information reports and newsletters*. Retrieved March 31, 2017, from www.dawsoncreek.ca/water

Conservation Ontario. (2017). *Integrated watershed management*. Retrieved January 5, 2018, from http://conservationontario.ca/policy-priorities/integrated-watershed-management/

Council of Canadian Academies. (2014). *Environmental impacts of shale gas extraction in Canada: The expert panel on harnessing science and technology to understand the environmental impacts of shale gas extraction*. Ottawa, Canada: Council of Canadian Academies. Retrieved March 31, 2017, from http://www.scienceadvice.ca/en/assessments/completed/shale-gas.aspx

Council of Canadians. (2012). News: Shell buys 10-years of water with funding of town's wastewater plant water. Edmonton, Alberta, November 10, 2012. Retrieved March 31, 2017, from http://canadians.org/fr/node/8933

Environment Canada. (2001). *Threats to sources of drinking water and aquatic ecosystem health in Canada* (National Water Research Institute, Scientific Assessment Report Series No. 1). Ottawa: Environment Canada.

Environmental Law Centre. (2015). *Professional reliance and environmental regulation in British Columbia*. Victoria, BC: Univeristy of Victoria, Faculty of Law. Retrieved March 31, 2017, from http://www.elc.uvic.ca/publications/professional-reliance-and-environmental-regulation-in-british-columbia/

FitzGibbon, J., Mitchell, B., & Veale, B. (2006). *Sustainable water management: State of practice in Canada and beyond*. Ottawa, ON: Canadian Water Resources Association.

Forest Practices Board. (2011a). *Cumulative effects: From assessment towards management* (Special Report FPB/SR/39). Forest Practices Board. Retrieved March 31, 2017, from https://www.bcfpb.ca/reports-publications/reports/cumulative-effects-from-assessment-towards-management/

Forest Practices Board. (2011b). *Cumulative effects assessment: A case study for the Kiskatinaw River Watershed* (Special Report FPB/SR/39). Victoria Forest Practices Board. Retrieved March 31, 2017, from http://www.dawsoncreek.ca/wordpress/wp-content/uploads/2011/08/SR39_CEA_Case_Study_for_the_Kiskatinaw_River_Watershed.pdf

Fraser Basin Council. (2013). *Workshop on water issues in BC's Peace Region*. Retrieved January 6, 2018, from https://www.fraserbasin.bc.ca/_Library/Water/ws_report_peace_water_issues_may-2013.pdf

GeoScience. (2016). *SkyTEM survey: British Columbia, Canada*. Data report by SkyTEM Surveys ApS. GeoScience BC Report 2016-03 retrieved Jan 6, 2018 from http://prrd.bc.ca/geoscience-bc-report-2016-03/

GeoScience BC. (n.d.). Montney water project. Retrieved March 31, 2017, from http://www.geosciencebc.com/s/Montney.asp,

Government of British Columbia. (1991). *Kiskatinaw river integrated watershed management plan, Victoria ministry of forests and ministry of environment*. Retrieved July 12, 2017, from http://www.dawsoncreek.ca/wordpress/wp-content/uploads/background-watershed-management-plans/Kiskatinaw_IMP_1991.pdf

Government of British Columbia. (1998). *Recommended Dawson Creek land and resource management plan*. Victoria, BC: British Columbia Land Use Coordination Office.

Government of British Columbia (2004). *Assessment of the city of Dawson Creek drinking water supply: Source water characteristics*, prepared by J. Jacklin. Prince george, BC: Ministry of Environment.

Government of British Columbia. (2007). *Kiskatinaw watershed water quality assessment and management*, prepared by G. Matscha & C. Van Geloven. Prince George, BC: Ministry of Environment.

Government of British Columbia. (2008). *Bacteria and parasite source identification in the Kiskatinaw river watershed*, prepared by G. Matsha. Prince George, BC: Ministry of Environment.

Government of British Columbia. (2014). Cumulative effects assessment for the south peace region operational trial version 2.3 report of the BC. Victoria, BC: Ministry of Environment. Retrieved July 12, 2017 from http://muskwa-kechika.com/uploads/documents/miscellaneous/M-KAB%20

temporary%20documents/Cumulative%20Effects%20Assessment%20for%20the%20South%20
Peace%20v2%203%20final.pdf

Government of British Columbia. (2015). Northeast water strategy: Ensuring the responsible use and management of Northeast British Columbia's water resources. Victoria, BC. Retrieved July 12, 2017, from http://www2.gov.bc.ca/assets/gov/environment/air-land-water/water/northeast-water-strategy/2015-northeast-water-strategy.pdf

Government of British Columbia. (2016). Water Sustainability Act. Retrieved January 10, 2018 from https://www2.gov.bc.ca/gov/content/environment/air-land-water/water/laws-rules/water-sustainability-act

Government of Canada. (2006). *Report of the expert panel on safe drinking water for first nations*. Ottawa, ON: Indian and Northern Affairs Canada. Retrieved March 31, 2017, from http://publications.gc.ca/site/eng/298371/publication.html

GW Solutions. (2012). *Montney Water Project - hydrogeological review*. Commissioned report prepared for the City of Dawson Creek retrieved Jan 6, 2018, from http://www.dawsoncreek.ca/wordpress/wp-content/uploads/watershed/GWS-final-Kiskatinaw-Source-Protection.pdf

Holmes, W., Dr. 2016. Personal communications and on-line.

InterraPlan Inc. (2014). "Towards a North-East BC Water Strategy: We Drink the Waters of the Peace – It is Our Life". A gathering of input First Nations opinions concerning water stewardship in the Treaty 8 Region of BC. Unpublished report for the Treaty 8 Tribal Association.

Intrinsik Environmental Science. (2014). *Detailed human health risk assessment of oil and gas activities in Northeastern British Columbia* (Report Prepared for BC Ministry of Health. Phase II. Report 10710). Calgary, Alberta. Retrieved March 31, 2017, from http://www.health.gov.bc.ca/library/publications/year/2014/health-risk-assessment-phase-two-recommendations.pdf

Laidlaw, D., & Passelac-Ross, M. (2010). Water rights and water stewardship: What about aboriginal peoples? Calgary, Alberta: Article in University of Calgary, Faculty of Law. Retrieved July 12, 2017, from https://ablawg.ca/wp-content/uploads/2010/07/blog_dl_mpr_waterrights_july2010.pdf

Lammers, R., Proussevitch, A. A., Frolking, A., & Grogan, D. S. (2013). Moving water on a malleable planet - Large scale inter-basin hydrological transfers now and in the future, American geophysical union, fall meeting, 2012. Retrieved March 31, 2017, from http://adsabs.harvard.edu/abs/2012AGUFMGC31D..03L

Lapp, S., & Whiten, R. (2012). *British Columbia/Alberta partnership for applied long-term watershed management research in the peace river region's upper Kiskatinaw river, Kamloops (BC): Forum for research and extension in natural resources*. Kamloops, BC: FORREX.

Mitchell, B., & Shrubsole D. (2007). An overview of integration in resource and environmental management. In K. Hanna & S. Slocombe (Eds.) *Integrated resource and environmental management: Concepts and practice* (pp. 21–35). Toronto: Oxford University Press.

Nelson, J. (2017). Site C and NAWAPA in the watershed sentinal. Retrieved from https://watershedsentinel.ca/articles/site-c-nawapa/

Parfitt, B. (2017a). *Fracking, first nations and water* (Report prepared for the Canadian Centre for Policy Alternatives). Vancouver, BC. Retrieved from https://www.policyalternatives.ca/protect-shared-waters

Parfitt, B. (2017b). A dam big problem report prepared for the Canadian centre for policy alternatives. Vancouver, BC. Retrieved from https://www.policyalternatives.ca/publications/reports/dam-big-problem

Paul, S., & Lee, J. (2013). Examining present and future water resources for the Kiskatinaw river – Final report on land-use and land cover change analysis. Prince George, BC: University of Northern British Columbia.

Pringle, C. (2001). Hydrologic connectivity and the management of biological reserves: A global perspective. *Ecological Applications, 11*(4), 981–998. doi:10.1890/1051-0761(2001)011[0981:HCATMO]2.0.CO;2

Saha G., & Lee, J. (2012). *Examining present and future water resources for the Kiskatinaw river – Final report on ground-water surface interactions and surface water quality sampling*, Prepared by Prince George, BC: University of Northern British Columbia.

Saulteau First Nations. (2015). *Saulteau first nations – Government of British Columbia new relationship and reconciliation agreement.* Victoria, BC: Queens Printer Publishing.

Shrubsole, D., Waters, D., Veale, B., & Mitchell, B. (2017). Integrated water management in Canada: The experience of watershed agencies. *International Journal of Water Resources Development 33*(3), 349–359. doi:10.1080/07900627.2016.1244048

Supreme Court of Canada. (2017). First Nation of Nacho Nyak Dun, et al. v. Government of Yukon. Retrieved Dec 3, 2017 from http://www.scc-csc.ca/case-dossier/info/sum-som-eng.aspx?cas=36779

Whiten, R. (2012). *Watershed program annual report*, Prepared by for City of Dawson Creek: Moberly Lake, InterraPlan.

Whiten, R. (2013). *Watershed program annual report*, Prepared by for City of Dawson Creek. Moberly Lake: InterraPlan.

Index

INDEX

Printed and bound by CPI Group (UK) Ltd, Croydon, CR0 4YY

24/10/2024

01778294-0004